AF517873

Practices and Techniques of Climate Smart Agriculture

Practices and Techniques of Climate Smart Agriculture

– Editors –

Dr. Rashmi Nigam

*Assistant Professor,
Department of Plant Pathology, J.V. College, Baraut, Uttar Pradesh*

Dr. Joginder Singh

*Assistant Professor,
Department of Horticulture, J.V. College, Baraut, Uttar Pradesh*

Prof. Sanjay Swami

SNRM, CPGSAS, (CAU), UMIAM (Barapani) Meghalaya

Mr. Mohit Husain

Scientist, FRI, Dehradun, Uttrakhand

Dr. Hemant Kumar

Assistant Professor, College of Forestry, SHUATS, Prayagraj, Uttar Pradesh

Dr. Rajesh Jatav

*Department of Horticulture,
College of Agriculture, Race Course Road, Gwalior, Madhya Pradesh*

BIOTECH

2022

BIOTECH BOOKS®

© 2022 EDITORS
ISBN 978-81-7622-506-9

All rights reserved. Including the right to translate or to reproduce this book or parts thereof except for brief quotations in critical reviews.

The authors are solely responsible for the contents of the bock chapters compiled in this book. The editors or publisher do not take any, responsibility for same in any manner. Errors, if any are purely unintentional and readers are requested to communicate such errors to the editors or publisher to avoid discrepancies in future.

Published by: **BIOTECH BOOKS**®
4762-63/23, Ansari Road, Darya Ganj,
New Delhi - 110 002
Phone: 09899295259, 011-43003222
E-mail: biotechbooks@yahoo.co.in
Website: www.biotechbooks.in

Typeset at:**Classic Computer Services**
Delhi - 110 035

Printed at: **Replika Press Pvt. Ltd.**

PRINTED IN INDIA

Preface

Climate-smart agriculture is an approach for reorienting agricultural systems to support food security under the new realities of climate change. Widespread changes in climate like rainfall and temperature patterns threaten agricultural production and increase the vulnerability of people dependent on agriculture for their livelihoods. Climate change disrupts food markets, posing population-wide risks to food supply. Climate Smart Agriculture is agriculture that sustainably increase productivity, resilience, reduces Greenhouse gases and enhances the achievement of national food security and development goal's. Climate Agriculture is an approach for developing agriculture strategies to secure sustainable food security under climate change. Climate smart agriculture provide the means to help stake holders from local to national and international levels identify agriculture strategies suitable to their local conditions.

The Book *"Practices and Techniques of Climate Smart Agriculture"* presents updates on climate smart agriculture for sustainable agriculture which have contributed significantly to increase agriculture production. The book discusses the emerging techniques in analyzing the vast resources of information regarding smart agriculture and its practical application and utilization. The book is structured into various chapters and primarily for the degree, post graduate students and for the researchers. Written by different authors this book provide in depth analysis and a frame work of new research. I wish to express our sincere thanks and deep sense of gratitude and indebtness to those who helped us directly or indirectly during the preparation of the manuscript of this text. I especially thankful our co-editors who have helped with me in editing the book.

Dr. Rashmi Nigam

Dr. Joginder Singh

Prof. Sanjay Swami

Mr. Mohit Husain

Dr. Hemant Kumar

Dr. Rajesh Jatav

Contents

Chapter 1

History and Evidence of Climate Change

B. Rajendra Kumar[1] *and L. Suryanarayana*[2]

[1]*Assistant Professor, Department of Agronomy, Agricultural College, Naira, Acharya N.G. Ranga Agricultural University, Andhra Pradesh*

[2]*Principal Scientist, Genetics and Plant Breeding, Agricultural Research Station, Amadalavalasa, Acharya N.G. Ranga Agricultural University, Andhra Pradesh*

The Earth's climate have continued to evolve over recent decades, increasing evidence of anthropogenic influences on climate change has been found. Correspondingly, the IPCC has made increasingly more definitive statements about human impacts on climate. It is now more certain than ever, based on many lines of evidence, that humans are changing Earth's climate. The atmosphere and oceans have warmed, accompanied by sea-level rise, a strong decline in Arctic sea ice, and other climate-related changes. The evidence is clear. However, due to the nature of science, not every single detail is ever totally settled or completely certain. Scientific evidence continues to be gathered around the world, and assumptions and findings about climate change are continually analysed and tested. Some areas of active debate and ongoing research include the link between ocean heat content and the rate of warming, estimates of how much warming to expect in the future, and the connections between climate change and extreme weather events. Awareness and a partial understanding of most of the interactive processes in the Earth system that govern climate and climate change predate the IPCC, often by many decades. A deeper understanding and quantification of these processes and their incorporation in climate models have progressed rapidly since the IPCC First Assessment Report in 1990.

Keywords: *Greenhouse gases, Climate change, Aerosols, Emissions and Global warming.*

Introduction

The history of the scientific discovery of climate change began in the early 19th century when ice ages and other natural changes in paleoclimate were first suspected and the natural greenhouse effect first identified. In the late 19th century, scientists first argued that human emissions of greenhouse gases could change the climate. Many other theories of climate change were advanced, involving forces from volcanism to solar variation. In the 1960s, the warming effect of carbon dioxide gas became increasingly convincing. Some scientists also pointed out those human activities that generated atmospheric aerosols could have cooling effects as well. During the 1970s, scientific opinion increasingly favored the warming viewpoint. Greenhouse gases were deeply involved in most climate changes and human caused emissions were bringing discernible global warming. Since the 1990s, scientific research on climate change has included multiple disciplines and has expanded. Climate change is caused by factors that include oceanic processes, biotic processes, variations in solar radiation received by Earth, plate tectonics and volcanic eruptions, and human-induced alterations of the natural world. The latter effect is currently causing global warming, and "climate change" is often used to describe human-specific impacts.

History

We are living in what is scientifically known as an 'interglacial' period, which is often thought of as a 'brief summer' that has lasted for about the past 10,000years. These brief warm periods have occurred about every 90,000 years, so that there are known to have been 9 ice ages in the past 780,000 years. Scientists have examined deep ocean sediment cores and ice cores from Greenland and the Antarctic and they all provide similar evidence. For instance, the Vostock ice core taken from the Antarctic shows 4 glacial periods in the past 420,000 years.

The planet's temperature at the coldest time during the last glacial period was up to 10°c colder than the present, but the average temperature in the middle of the last 'ice age' was only about 5°c below today's average temperature. The difference now is that our burning of 'fossil fuels' has caused global average temperatures to rise by about 0.7°c in the past 100years. Global climate models predict a rise in global average temperatures between 1990 and 2100 of between 1.5°C and 5.8°C, depending on future carbon dioxide emissions. However, the climate is not stable even without the greenhouses gases added by man. When the weather became colder at the end of each 'interglacial period', water became trapped in vast sheets of ice covering much of the Northern Hemisphere, that extended as far south as southern England. As water was locked up in land ice, sea levels fell by as much as 120 metres.

Temperatures during the interglacial periods varied, they were not always as they are today. In some periods the temperature was higher, resulting in greater glacial melting and higher sea levels, while in others the temperature was lower, less ice melted and the sea level was lower. For instance, it is thought that the temperature during the last interglacial period was slightly warmer than today and that this caused greater melting of the Greenland glaciers, resulting in a sea level that was several metres higher than the present

To confuse us further, the level of the land surface has been rising throughout this period, caused by an uplifting of the European Tectonic Plate as Africa collided with Southern Europe. Evidence of the higher sea levels can be seen in the wave cut platforms, fossil cliffs, cliff notches, and raised beaches that can be seen about 8 metres, 18 metres and 30 metres above current mean sea level. The wave cut platforms and raised beaches that are about 3 metres above the current high water level were created in the last 'interglacial' warm period about 120,000 years ago. There is a large raised beach area of18 metres and 30 metres above sea level. Each of these beaches was created by the sea level during interglacial periods in the last 450,000 years. sea levels rose 50 metres between 11,000 and 6,000 years ago to cut our islands.

Regional Changes, Antiquity from 19th Century

From ancient times, people suspected that the climate of a region could change over the course of centuries. Renaissance and later scholars saw that deforestation, irrigation, and grazing had altered the lands around the Mediterranean since ancient times. The most striking change came in the 18th and 19th centuries, within a single lifetime. The conversion of Eastern North America from forest to croplands. Meanwhile, national weather agencies had begun to compile masses of reliable observations of temperature and rainfall. By the end of the 19th century, scientific opinion had turned decisively against any belief in a human influence on climate.

Paleoclimate Change and Theories of its Causes, 19th Century

Prior to the 18th century, scientists had not suspected that prehistoric climates were different from the modern period. By the late 18th century, geologists found evidence of a succession of geological ages with changes in climate. Agassiz developed a theory of what he termed "Ice Age" William Buckland had led attempts in Britain to adapt the geological theory of catastrophism.

First Calculations of Human-Induced Climate Change, 1896

In 1896 Svante Arrhenius calculated the effect of a doubling atmospheric carbon dioxide to be an increase in surface temperatures of 5-6 degrees Celsius. By the late 1890s, American scientist Samuel Pier point Langley had attempted to determine the surface temperature of the Moon by measuring infrared radiation leaving the Moon and reaching the Earth. The angle of the Moon in the sky when a scientist took a measurement determined how much CO_2 and water vapor the Moon's radiation had to pass through to reach the Earth's surface, resulting in weaker measurements when the Moon was low in the sky. This result was unsurprising given that scientists had known about infrared radiation absorption for decades.

A Swedish scientist, Svante Arrhenius, used Langley's observations of increased infrared absorption where Moon rays pass through the atmosphere at a low angle, encountering more carbon dioxide, to estimate an atmospheric cooling effect from a future decrease of CO_2. He realized that the cooler atmosphere would hold less water vapor (another greenhouse gas) and calculated the additional cooling effect. He also realized the cooling would increase snow and ice cover at high latitudes, making the planet reflect more sunlight and thus further cool down, as James Croll had hypothesized. Overall Arrhenius calculated that cutting CO_2 in half would suffice to produce an ice age. He further calculated that a doubling of atmospheric CO_2 would give a total warming of 5-6 degrees Celsius. Further, Arrhenius colleague Professor Arvid Högbom, who was quoted in length in Arrhenius 1896 study on the Influence of Carbonic Acid in the Air upon the temperature of the earth had been attempting to quantify natural sources of emissions of CO_2 for purposes of understanding the global carbon cycle.

Högbom found that estimated carbon production from industrial sources in the 1890s (mainly coal burning) was comparable with the natural sources. Arrhenius saw that this human emission of carbon would eventually lead to warming. However, because of the relatively low rate of CO_2 production in 1896, Arrhenius thought the warming would take thousands of years, and he expected it would be beneficial to humanity.

Paleoclimates and Sunspots, Early 1900s to 1950s

Arrhenius's calculations were disputed and subsumed into a larger debate over whether atmospheric changes had caused the ice ages. Experimental attempts to measure infrared absorption in the laboratory seemed to show little differences resulted from increasing CO_2 levels, and also found significant overlap between absorption by CO_2 and absorption by water vapor, all of which suggested that increasing carbon dioxide emissions would have little climatic effect. These early experiments were later found to be insufficiently accurate, given the instrumentation of the time. Many scientists also thought that the oceans would quickly absorb any excess carbon dioxide. Other theories of the causes of climate change fared no better. The principal advances were in observational paleoclimatology, as scientists in various fields of geology worked out methods to reveal ancient climates. Wilmot H. Bradley found that annual varves of clay laid down in lake beds showed climate cycles. An Arizona astronomer, Andrew Ellicott Douglass, saw strong indications of climate change in tree rings. Noting that the rings were thinner in dry years, he reported climate effects from solar variations, particularly in connection with the 17th-century dearth of sunspots (the Maunder Minimum) noticed previously by William Herschel and others. Other scientists, however, found good reason to doubt that tree rings could reveal anything beyond random regional variations. The value of tree rings for climate study was not solidly established until the 1960s. Through the 1930s the most persistent advocate of a solar-climate connection was astrophysicist Charles Greeley Abbot. By the early 1920s, he had concluded that the solar "constant" was misnamed: his observations showed large variations, which he connected with sunspots passing across the face of the Sun. He and a few others pursued the topic into the 1960s, convinced that sunspot variations were a main cause of climate change. Other scientists were skeptical. Nevertheless, attempts to connect the solar cycle with climate cycles were popular in the 1920s and 1930s. Respected scientists announced correlations that they insisted were reliable enough to make predictions. Sooner or later, every prediction failed, and the subject fell into disrepute.

Meanwhile, the Serbian engineer Milutin Milankovitch, building on James Croll's theory, improved the tedious calculations of the varying distances and angles of the Sun's radiation as the Sun and Moon gradually perturbed the Earth's orbit. Some observations of varves (layers seen in the mud covering the bottom of lakes) matched the prediction of a Milankovitch cycle lasting about 21,000 years. However, most geologists dismissed the astronomical theory. For they could not fit Milankovitch's timing to the accepted sequence, which had only four ice ages, all of them much longer than 21,000 years. In 1938 a British engineer, Guy Stewart Callendar, attempted to revive Arrhenius's greenhouse-effect theory. Callendar presented evidence that both temperature and the CO_2 level in the atmosphere had been rising over the past half-century, and he argued that newer spectroscopic measurements showed that the gas was effective in absorbing infrared in the atmosphere. Nevertheless, most scientific opinion continued to dispute or ignore the theory.

Increasing Concern, 1950s-1960s

Better spectrography in the 1950s showed that CO_2 and water vapor absorption lines did not overlap completely. Climatologists also realized that little water vapor was present in the upper atmosphere. Both developments showed that the CO_2 greenhouse effect would not be overwhelmed by water vapor. In 1955 Hans Suess's carbon-14 isotope analysis showed that CO_2 released from fossil fuels was not immediately absorbed by the ocean. In 1957, better understanding of ocean chemistry led Roger Revelle to a realization that the ocean surface layer had limited ability to absorb carbon dioxide. Also predicting the rise in levels of CO_2, later being proven by Charles David Keeling. By the late 1950s, more scientists were arguing that carbon dioxide emissions could be a problem, with some projecting in 1959 that CO_2 would rise 25 per cent by the year 2000, with potentially "radical" effects on climate. In 1960 Charles David Keeling demonstrated that the level of CO_2 in the atmosphere was in fact rising. Concern mounted year by year along with the rise of the "Keeling Curve" of atmospheric CO_2.

Another clue to the nature of climate change came in the mid-1960s from analysis of deep-sea cores by Cesare Emiliani and analysis of ancient corals by Wallace Broecker and collaborators. Rather than four long ice ages, they found a large number of shorter ones in a regular sequence. It appeared that the timing of ice ages was set by the small orbital shifts of the Milankovitch cycles. While the matter remained controversial, some began to suggest that the climate system is sensitive to small changes and can readily be flipped from a stable state into a different one. Scientists meanwhile began using computers to develop more sophisticated versions of Arrhenius's calculations. In 1967, taking advantage of the ability of digital computers to integrate absorption curves numerically, Syukuro Manabe and Richard Wetherald made the first detailed calculation of the greenhouse effect incorporating convection (the "Manabe-Wetherald one-dimensional radiative-convective model"). They found that, in the absence of unknown feedbacks such as changes in clouds, a doubling of carbon dioxide from the current level would result in approximately 2 °C increase in global temperature. By the 1960s, aerosol pollution ("smog") had become a serious local problem in many cities, and some scientists began to consider whether the cooling effect of particulate pollution could affect global temperatures. Scientists were unsure whether the cooling effect of particulate pollution or warming effect of greenhouse gas emissions would predominate, but regardless, began to suspect that human emissions could be disruptive to climate in the 21st century if not sooner. In his 1968 book *The Population Bomb*, Paul R. Ehrlich wrote, "the greenhouse effect is being enhanced now by the greatly increased level of carbon dioxide, this is being countered by low-level clouds generated by contrails, dust, and other contaminants. At the moment we cannot predict what the overall climatic results will be of our using the atmosphere as a garbage dump.

In 1969, NATO was the first candidate to deal with climate change on an international level. It was planned then to establish a hub of research and initiatives of the organization in the civil area, dealing with environmental topics as Acid Rain and the Greenhouse effect. The suggestion of US President Richard Nixon was not very successful with the administration of German Chancellor Kurt Georg Kiesinger. But the topics and the preparation work done on the NATO proposal by the German authorities gained international momentum, (see *e.g.*, the Stockholm United Nations Conference on the Human Environment 1970) as the government of Willy Brandt started to apply them on the civil sphere instead.

Scientists Increasingly Predict Warming, 1970s

Mean temperature anomalies during the period 1965 to 1975 with respect to the average temperatures from 1937 to 1946. This dataset was not available at the time. In the early 1970s, evidence that aerosols were increasing world-wide encouraged Reid Bryson and some others to warn of the possibility of severe cooling. Meanwhile, the new evidence that the timing of ice ages was set by predictable orbital cycles suggested that the climate would gradually cool, over thousands of years. For the century ahead, however, a survey of the scientific literature from 1965 to 1979 found 7 articles predicting cooling and 44 predicting warming (many other articles on climate made no prediction); the warming articles were cited much more often in subsequent scientific literature. Several scientific panels from this time period concluded that more research was needed to determine whether warming or cooling was likely, indicating that the trend in the scientific literature had not yet become a consensus.

John Sawyer published the study *Man-made Carbon Dioxide and the "Greenhouse" Effect* in 1972. He summarized the knowledge of the science at the time, the anthropogenic attribution of the carbon dioxide greenhouse gas, distribution and exponential rise, findings which still hold today. Additionally he accurately predicted the rate of global warming for the period between 1972 and 2000. *The increase of 25 per cent CO_2 expected by the end of the century therefore corresponds to an increase of 0.6°C in the world temperature – an amount somewhat greater than the climatic variation of recent centuries*. - John Sawyer, 1972. The mainstream news media at the time exaggerated the warnings of the minority who expected imminent cooling. For example, in 1975, *Newsweek* magazine published a story that warned of "ominous signs that the Earth's weather patterns have begun to change." The article continued by stating that evidence of global cooling was so strong that meteorologists were having "a hard time keeping up with it." On October 23, 2006, *Newsweek* issued an update stating that it had been "spectacularly wrong about the near-term future". In the first two "Reports for the Club of Rome" in 1972and 1974, the anthropogenic climate changes by CO_2 increase as well as by Waste heat were mentioned. About the latter John Holdren wrote in a study cited in the 1st report, *"that global thermal pollution is hardly our most immediate environmental threat. It could prove to be the most inexorable, however, if we are fortunate enough to evade all the rest."* Simple global-scale estimates that recently have been actualized and confirmed by more refined model calculations show noticeable contributions from waste heat to global warming after the year 2100, if its growth rates are not strongly reduced (below the averaged 2 per cent p.a. which occurred since 1973). Evidence for warming accumulated. By 1975, Manabe and Wetherald had developed a three-dimensional Global climate model that gave a roughly accurate representation of the current climate. Doubling CO_2 in the model's atmosphere gave a roughly 2 °C rise in global temperature. Several other kinds of computer models gave similar results: it was impossible to make a model that gave something resembling the actual climate and not have the temperature rise when the CO_2 concentration was increased.

In a separate development, an analysis of deep-sea cores published in 1976 by Nicholas Shackleton and colleagues showed that the dominating influence on ice age timing came from a 100,000-year Milankovitch orbital change. This was unexpected, since the change in sunlight in that cycle was slight. The result emphasized that the climate system is driven by feedbacks, and thus is strongly susceptible to small changes in conditions. In July 1979 the United States National Research Council published a report, concluding (in part): When it is assumed that the CO_2 content of the atmosphere is doubled and statistical thermal equilibrium is achieved, the more realistic of the modeling efforts predict a global surface warming of between 2°C and 3.5°C, with greater increases at high latitudes. We have tried but have been unable to find any overlooked or underestimated physical effects that could reduce the currently estimated global warming due to a doubling of atmospheric CO_2 to negligible proportions or reverse them altogether. The 1979 World Climate Conference of the World Meteorological Organization concluded "it appears plausible that an increased amount of carbon dioxide in the atmosphere can contribute to a gradual warming of the lower atmosphere, especially at higher latitudes. It is possible that some effects on a regional and global scale may be detectable before the end of this century and become significant before the middle of the next century.

Consensus Begins to Form, 1980-1988

By the early 1980s, the slight cooling trend from 1945-1975 had stopped. Aerosol pollution had decreased in many areas due to environmental legislation and changes in fuel use, and it became clear that the cooling effect from aerosols was not going to increase substantially while carbon dioxide levels were progressively increasing. In 1982, Greenland ice cores drilled by Hans Oeschger, Willi Dansgaard, and collaborators revealed dramatic temperature oscillations in the space of a century in the distant past.[51] The most prominent of the changes in their record corresponded to the violent Younger Dryas climate oscillation seen in shifts in types of pollen in lake beds all over Europe. Evidently drastic climate changes were possible within a human lifetime. James Hansen during his 1988 testimony to Congress, which alerted the public to the dangers of global warming. In 1973, British scientist James

Lovelock speculated that chlorofluorocarbons (CFCs) could have a global warming effect. In 1975, V. Ramanathan found that a CFC molecule could be 10,000 times more effective in absorbing infrared radiation than a carbon dioxide molecule, making CFCs potentially important despite their very low concentrations in the atmosphere. While most early work on CFCs focused on their role in ozone depletion, by 1985 Ramanathan and others showed that CFCs together with methane and other trace gases could have nearly as important a climate effect as increases in CO_2. In other words, global warming would arrive twice as fast as had been expected. In 1985 a joint UNEP/WMO/ICSU Conference on the "Assessment of the Role of Carbon Dioxide and Other Greenhouse Gases in Climate Variations and Associated Impacts" concluded that greenhouse gases "are expected" to cause significant warming in the next century and that some warming is inevitable. Meanwhile, ice cores drilled by a Franco-Soviet team at the Vostok Station in Antarctica showed that CO_2 and temperature had gone up and down together in wide swings through past ice ages. This confirmed the CO_2-temperature relationship in a manner entirely independent of computer climate models, strongly reinforcing the emerging scientific consensus. The findings also pointed to powerful biological and geochemical feedbacks.

In June 1984, James E. Hansen made one of the first assessments that human-caused warming had already measurably affected global climate. Shortly after, a "World Conference on the Changing Atmosphere: Implications for Global Security" gathered hundreds of scientists and others in Toronto. They concluded that the changes in the atmosphere due to human pollution "represent a major threat to international security and are already having harmful consequences over many parts of the globe," and declared that by 2005 the world should push its emissions some 20 per cent below the 1988 level. The 1980s saw important breakthroughs with regard to global environmental challenges *e.g.*, Ozone depletion was mitigated by the Vienna Convention (1985) and the Montreal Protocol (1987). Acid rain was mainly regulated on the national and regional level. 2015 – Warmest Global Year on Record (since 1880) – Colors indicate temperature anomalies (NASA/NOAA; 20 January, 2016).

In 1988 the WMO established the Intergovernmental Panel on Climate Change with the support of the UNEP. The IPCC continues its work through the present day, and issues a series of Assessment Reports and supplemental reports that describe the state of scientific understanding at the time each report is prepared. Scientific developments during this period are summarized about once every five to six years in the IPCC Assessment Reports which were published in 1990 (First Assessment Report), 1995 (Second Assessment Report), 2001 (Third Assessment Report), 2007 (Fourth Assessment Report), and 2013/2014 (Fifth Assessment Report).[58]

Since the 1990s, research on climate change has expanded and grown, linking many fields such as atmospheric sciences, numeric modeling, behavioral sciences, geology and economics. Articles on climate change science are now frequently published in major journals such as Science by the American Association for the Advancement of Science and Nature, in addition there are focused journals on climate change research such as Nature Climate Change, Climate Change, Journal of Climate, Wiley Interdisciplinary Reviews: Climate Change, and International Journal of Climate Change Strategies and Management; furthermore, many journals on related subjects continue to publish articles that build out the science behind climate change (*e.g.* Quaternary Research).

Discovery of other Climate Changing Factors

Methane: In 1861, John Tyndall determined that coal gas, a mix of methane and other gases, strongly absorbed infrared radiation. Methane was subsequently detected in the atmosphere in 1948, and in the 1980s scientists realized that human emissions were having a substantial impact.[52]. Chlorofluorocarbon: In 1971, British scientist James Lovelock speculated that chlorofluorocarbons (CFCs) could have a global warming effect. In 1975, V. Ramanathan found that a CFC molecule could be 10,000 times more effective in absorbing infrared radiation than a carbon dioxide molecule, making CFCs potentially important despite their very low concentrations in the atmosphere. While most early work on CFCs focused on their role in ozone depletion, by 1985 scientists had concluded that CFCs together with methane and other trace gases could have nearly as important a climate effect as increases in CO_2.

Evidence for Climate Change

During the most recent history, scientists have been able to construct evidence of climate change from collected information on temperature, rainfall and other weather variables from measuring stations all over the world. The oldest time series of climate data is a temperature record from central England beginning in the 17th century. Most instrumental records however, date back only to the 19thcentury. Despite this relatively short period of data collection, by careful mathematical analysis scientists have been able to demonstrate that the Earth's surface has warmed on average by about 0.6 °C during the 20thcentury. Such warming, it is believed, is related largely to mankind's pollution of the atmosphere by greenhouse gases. It is now known that the Earth's climate has changed many times before mankind began altering the global environment. Before we began collecting information about the weather. Knowledge of palaeoclimate changes has come from the reconstruction of indirect or proxy sources of information. For climate changes that have occurred during the last 2,000 years.

Other forms of proxy data come from the study of natural phenomena which are climate-dependent. The growth of tree rings, for example, is dependent on climate, and tree ring sizes and wood densities may be used to reconstruct climates at the time of growth. The oldest surviving trees are about 4,000 years old. Older wood from dead trees has been used to reconstruct past climates back about 10,000 years. To reconstruct the climate of the last Ice Age, scientists have had to turn to the chemical and physical properties of ice, collected from ice cores drilled in Greenland and Antarctica. As snow is laid down and compacted it stores a record of the current climate. Palaeoclimatologists have also been able to use sea sediments to reconstruct climate during the last Ice Age and many before it, stretching back millions of years. To investigate the oldest climate changes. Some stretching back many hundreds of millions of years, geologists used clues from rocks. Such geological evidence has been used to confirm ideas that continental drift and mountain building can change the Earth's climate over hundreds of millions of years.

Rocks

Permian sandstone, coloured red with iron oxides, indicating a hot climate at the time of deposition. The characteristics of different rocks depend on the environment in which the sediments were deposited. Some sands and gravels are dropped by glaciers as they melt and they become a distinctive rock called till. Where till is found there must have been glaciers and therefore it must have been cold. Rocks that form in a hot desert environment are often coloured red with iron deposits. Also at high temperatures, sea water can evaporate quickly leaving behind a layer of salt on the ground which becomes preserved in the rocks and is another indicator of a hot climate.

Fossils

Coral reefs live in tropical waters. They need a particular temperature, a specific depth of water and the right amount of light. If the depth of the water changes just a fraction, they cannot survive. Different species of plants and animals need different conditions to survive. Some plants and animals can be very sensitive to climate and do not adapt easily to change. Therefore where fossil corals are found it is possible to estimate fairly precisely the environment they lived in by assuming that they needed the same conditions as those that thrive today.

Pollen and Spores

Schulzospora campyloptera a fossil plant spore from the Carboniferous Period. Plants produce pollens and spores that are particularly useful in helping to determine climate. They are tiny with a resistant outer case and are produced in millions. This means that they can be covered in mud quickly and are more easily preserved as fossils than large animals. Each plant has different shaped pollen or spores so when the fossil is put under a microscope it is possible to identify the type of plant it came from. Different plants are adapted to different climates therefore looking at all the types of pollen present in a layer of rock can be a good indication of the climate at the time when they were living.

Microfossils

Microfossils are tiny fossils, usually smaller than 4 mm in size. They can be the remains of minuscule life forms or microscopic parts of plants and animals, for example tiny bones from larger animals. Like other plants and animals they may require specific living conditions, but they are more abundant and are found in many types of sedimentary rocks. Because of this they can be a useful indicator of how climate has changed over a period of time. *Cytherelloidea*, a microscopic genus of ostracod, is found only in modern seas where the temperature does not fall below 10°C and this preference for warm water has been traced back into geological time.

During the early Cretaceous Period *Cytherelloidea* was confined to the Tethys Sea but species migrated northwards (Warm phases) on several occasions. Fossils are indicators of past climatic conditions.

Landscape

Today's landscape gives clues to the climate of the past. Glaciers for instance, left telltale signs of their activity. Glaciers also pick up pieces of rock and gravel and, as they move forwards, this debris scratches grooves called 'striations' into the rocks on the valley floor. When the ice melts, the very large pieces of rock they have been carrying are dumped, often many miles from where they were picked up. These abandoned boulders are called 'erratics'. All of these kinds of erosion and deposition, clearly tell of temperatures much colder than our current climate.

Oxygen Isotopes

Antarctic ice cores hold the secrets of our past climate. A European project revealed a climate record for the last 800 000 years. It showed that the Earth experience eight glacial cycles each with a warm period and an ice age. The oxygen isotope composition of the minerals is affected by the air temperature so soda straws may record a long history of changing conditions as they grow. There is more than one type of oxygen and these different types are called isotopes. Some of these isotopes are 'heavier' than others and this mass is determined by the number of neutrons the isotopes contain. Oxygen-16 (containing 8 protons and 8 neutrons in each atomic nucleus) is lighter than oxygen-18 (8 protons and 10 neutrons). In water the relative amount of each of type of oxygen varies with the temperature. Rain and ice contain water high in oxygen-16 so when there is lots of ice the ocean water is left with a lot of oxygen-18.

Shells and sediments carry the isotopic signature of the water (Oxygen-16) in which they were formed. Therefore, by analysing the ratio of oxygen-16 to oxygen-18 recorded in rocks, fossils, ice and sediments we can find evidence of the climate at the time they were formed. Water vapour gradually loses oxygen-18 as it travels from the equator to the poles. Because water molecules with heavy oxygen-18 isotopes in them condense more easily than normal water molecules, air becomes progressively depleted in oxygen-18 as it travels to high latitudes and becomes colder and drier. In turn, the snow that forms most glacial ice is also depleted in oxygen-18. As glacial ice melts, it returns oxygen-16 rich freshwater to the ocean. Therefore, oxygen isotopes preserved in ocean sediments provide evidence for past ice ages and records of salinity.

Climate Change

Climate change is defined as a change in the statistical distribution of weather patterns when that change lasts for an extended period of time (*i.e.*, decades to millions of years). Accordingly, fluctuations over periods shorter than a few decades, such as El Niño, do not represent climate change. The term sometimes is used to refer specifically to climate change caused by human activity and the term *climate change* has become synonymous with anthropogenic global warming. Climate change may refer to a change in average weather conditions, or in the time variation of weather around longer-term average conditions.

Climate change is caused by factors such as biotic processes, variations in solar radiation received by Earth, plate tectonics, and volcanic eruptions. Certain human activities have also been identified as significant causes of recent climate change, often referred to as *global warming*. Scientists actively work to understand past and future climate by using observations and theoretical models. A climate record extending deep into the Earth's past has been assembled, and continues to be built up, based on geological evidence from temperature profiles, ice, floral and faunal records, glacial and periglacial processes, stable-isotope and other analyses of sediment layers, and records of past sea levels.

Climatic Change Versus Climate Change

In 1966, the World Meteorological Organization (WMO) proposed the term climatic change to encompass all forms of climatic variability on time-scales longer than 10 years, whether the cause was natural or anthropogenic. The term climate change replaced climatic change as the dominant term to reflect an anthropogenic cause.

Causes

On the broadest scale, the rate at which energy is received from the Sun and the rate at which it is lost to space determine the equilibrium temperature and climate of Earth. This energy is distributed around the globe by winds, ocean currents, and other mechanisms to affect the climates of different regions. Factors that can shape climate are called climate forcings or "forcing mechanisms". These include processes such as variations in solar radiation, Earth's orbit, albedo or reflectivity of the continents and oceans, mountain-building and continental drift and changes in greenhouse gas concentrations. There are also key threshold factors which when exceeded can produce rapid change. Forcing mechanisms can be either "internal" or "external". Internal forcing mechanisms are natural processes within the climate system itself (*e.g.*, the thermohaline circulation). External forcing mechanisms can be either natural (*e.g.*, changes in solar output) or anthropogenic (*e.g.*, increased emissions of greenhouse gases).

Internal Forcing Mechanisms

Scientists generally define the five components of earth's climate system to include atmosphere, hydrosphere, cryosphere, lithosphere (restricted to the surface soils, rocks, and sediments), and biosphere. Natural changes in the climate system ("internal forcings") result in internal "climate variability". The ocean is a fundamental part of the climate system, some changes in it occurring at longer timescales than in the atmosphere, Short-term fluctuations (years to a few decades) such as the *El Niño*-Southern Oscillation, the Pacific decadal oscillation, the North Atlantic oscillation, and the Arctic oscillation, represent climate variability rather than climate change. Life affects climate through its role in the carbon and water cycles and through such mechanisms as albedo, evapotranspiration, cloud formation, and weathering.

External Forcing Mechanisms

Orbital Variations

Slight variations in Earth's orbit lead to changes in the seasonal distribution of sunlight reaching the Earth's surface and how it is distributed across the globe. The Sun is the predominant source of energy input to the Earth. Other sources include geothermal energy from the Earth's core, and heat from the decay of radioactive compounds. Both long- and short-term variations in solar intensity are known to affect global climate. Three to four billion years ago, the Sun emitted only 70 per cent as much power as it does today. If the atmospheric composition had been the same as today, liquid water should not have existed on Earth. However, there is evidence for the presence of water on the early Earth, what is known as the faint young Sun paradox. The Great Oxygenation Event oxygenation of the atmosphere around 2.4 billion years ago was the most notable alteration. Over the next five billion years, the Sun's ultimate death as it becomes a red giant and then a white dwarf will have large effects on climate, with the red giant phase possibly ending any life on Earth that survives until that time.

Volcanism

The eruptions considered to be large enough to affect the Earth's climate on a scale of more than 1 year are the ones that inject over 100,000 tons of SO_2 into the stratosphere. This is due to the optical properties of SO_2 and sulfate aerosols, which strongly absorb or scatter solar radiation, creating a global layer of sulfuric acid haze. On average, such eruptions occur several times per century, and cause cooling (by partially blocking the transmission of solar radiation to the Earth's surface) for a period of a few years. The eruption of Mount Pinatubo in 1991, the second largest terrestrial eruption of the 20th century, affected the climate substantially, subsequently global temperatures decreased by about 0.5°C (0.9°F) for up to three years. Thus, the cooling over large parts of the Earth reduced surface temperatures in 1991-93.

Plate Tectonics

Over the course of millions of years, the motion of tectonic plates reconfigures global land and ocean areas and generates topography. This can affect both global and local patterns of climate and atmosphere-ocean circulation. The size of continents is also important. Because of the stabilizing effect of the oceans on temperature, yearly temperature variations are generally lower in coastal areas than they are inland. A larger supercontinent will therefore have more area in which climate is strongly seasonal than will several smaller continents or islands.

Human Influences

In the context of climate variation, anthropogenic factors are human activities which affect the climate which" is largely irreversible. Of most concern in these anthropogenic factors is the increase in CO_2 levels due to emissions from fossil fuel combustion, followed by aerosols (particulate matter in the atmosphere) and the CO_2 released by cement manufacture. Other factors, including land use, ozone depletion, and deforestation, are affecting climate.

Evidence for climatic change is taken from a variety of sources:

- Temperature measurements and proxies
- Historical and archaeological evidence
- Glaciers
- Glaciers are considered among the most sensitive indicators of climate change. Their size is determined by a mass balance between snow input and melt output.
- Arctic sea ice loss

The decline in Arctic sea ice, both in extent and thickness, over the last several decades is further evidence for rapid climate change. Sea ice is frozen seawater that floats on the ocean surface. It covers millions of square miles in the polar regions, varying with the seasons. Satellite observations show that Arctic sea ice is now declining at a rate of 13.3 per cent per decade, relative to the 1981 to 2010 average.

Vegetation

A change in the type, distribution and coverage of vegetation may occur given a change in the climate. Some changes in climate may result in increased precipitation and warmth, resulting in improved plant growth and the subsequent sequestration of airborne CO_2. A gradual increase in warmth in a region will lead to earlier flowering and fruiting times, driving a change in the timing of life cycles of dependent organisms. Conversely, cold will cause plant bio-cycles to lag. Larger, faster or more radical changes, however, may result in vegetation stress, rapid plant loss and desertification in certain circumstances.

Pollen Analysis

Palynology is the study of contemporary and fossil palynomorphs, including pollen. Palynology is used to infer the geographical distribution of plant species, which vary under different climate conditions. Different groups of plants have pollen with distinctive shapes and surface textures, and since the outer surface of pollen is composed of a very resilient material, they resist decay. Changes in the type of pollen found in different layers of sediment in lakes, bogs, or river deltas indicate changes in plant communities. These changes are often a sign of a changing climate.

Cloud Cover and Precipitation

Past precipitation can be estimated in the modern era with the global network of precipitation gauges. Surface coverage over oceans and remote areas is relatively sparse. Satellite clouds and precipitation data has been available since the 1970s. Estimated global land precipitation increased by approximately 2 per cent over the course of the 20th century.

Dendroclimatology

Dendroclimatology is the analysis of tree ring growth patterns to determine past climate variations. Wide and thick rings indicate a fertile, well-watered growing period, whilst thin, narrow rings indicate a time of lower rainfall and less-than-ideal growing conditions.

Ice Cores

Analysis of ice in a core drilled from an ice sheet such as the Antarctic ice sheet, can be used to show a link between temperature and global sea level variations. The air trapped in bubbles in the ice can also reveal the CO_2 variations of the atmosphere provide valuable information about the differences between ancient and modern atmospheric conditions.

Animals

Remains of beetles are common in freshwater and land sediments. Different species of beetles tend to be found under different climatic conditions, knowledge of the present climatic range of the different species, and the age of the sediments in which remains are found, past climatic conditions may be inferred. Similarly, the historical abundance of various fish species has been found to have a substantial relationship with observed climatic conditions.

Sea Level Change

Global sea level change for much of the last century has generally been estimated using tide gauge measurements collated over long periods of time to give a long-term average have provided an improved measurement of global sea level change.

Conclusions

The physical mechanisms by which changes in the amounts of greenhouse gases cause climate changes. It discusses the evidence that the concentrations of these gases in the atmosphere have increased and are still increasing rapidly, that climate change is occurring, and that most of the recent change is almost certainly due to emissions of greenhouse gases caused by human activities. The history of the centuries, long effort to document and understand climate change is often complex, marked by successes and failures, and has followed a very uneven pace. Climate change science is now contributing to the foundation of a new interdisciplinary approach to understanding our environment.

REFERENCES

Hansen, J., *et al.*, 1984: Climate sensitivity: Analysis of feedback mechanisms. In: *Climate Processes and Climate Sensitivity* [Hansen, J.E., and T. Takahashi (eds.)]. Geophysical Monograph 29, American Geophysical Union, Washington, DC, pp. 130–163.

IPCC, 1990: *Climate Change: The IPCC Scientific Assessment* [Houghton, J.T., G.J. Jenkins, and J.J. Ephraums (eds.)].

Cambridge University Press, Cambridge, United Kingdom and New York, NY, USA, 365 pp.

Lovelock, J.E., 1971: Atmospheric fluorine compounds as indicators of air movements. *Nature*, 230, 379–381.

Ramanathan, V., 1975: Greenhouse effect due to chloro fluoro carbons: Climatic implications. *Science*, 190, 50–52.

Tyndall, J., 1861: On the absorption and radiation of heat by gases and vapours, and on the physical connection, *Philos. Mag.*, 22, 277–302.

Chapter 2

Climate Smart Agriculture

J. Anusha[1]*, *M. Arun Kumar*[2] *and Ch. Ramya*[1]

[1]*Ph.D Scholar, Department of Agricultural Economics, PJTSAU, Hyderabad – 500 030, Telangana*

[2]*Ph.D Scholar, Department of Agricultural Economics, BCKV, Mohanpur, West Bengal*

Change is nature's law, it's inevitable, if it is by the virtue of the nature is welcome. World climate is changing day by day, it affects the life cycles on the earth. Agriculture is also prone to the climate change which is directly and indirectly affecting the farming practices and crop yield. Agriculture field faces many problems due to climate change as well as population; it will be 9.1 Billion on 2050.To feed this population we need to take sustainable approach. As, food and agriculture could be game changers in 2030. Climate Smart Agriculture is agriculture that sustainably increase productivity, resilience, reduces Greenhouse Gases, enhances the achievement of national food security and development goal's. Climate Smart Agriculture adopts many practices and technologies to make agriculture production system most efficient and resilience. Climate smart agriculture is effective but difficult to adapt immediately. It requires special skill, knowledge, collaboration of people and requires several years for proper implementation.

Keywords: Climate smart agriculture, Climate change, Production, Resilience.

Introduction

Climate Smart Agriculture (CSA)

Climate Smart Agriculture is agriculture that sustainably increase productivity, resilience, reduces Greenhouse Gases and enhances the achievement of national food security and development goal's (FAO, 2010).The Food and Agriculture Organisation of world has initiated this concept. Change is nature's law, it's inevitable, if it is by the virtue of the nature is welcome. Agriculture also prone to the climate change which is directly and indirectly affecting the farming practices and crop yield. Agriculture field faces many problems due to climate change. World population increasing day by day, it will be 9.1 Billion on 2050. To feed this much of the population agriculture production must have to be increase by 60 per cent to meet the future requirement. There is high demand for adoption of climate smart agriculture (Sustainable Agriculture).

Why is Climate Smart Agriculture (CSA) Needed?

- As per FAO estimate, by the year 2050 world population will increase by 1/3 rd and food required for food security will increased by 60 per cent. Since, last decade climate change has cumulative impact effect on productivity. Agriculture becomes a high risk profession for farmers increasingly like better to migrate. In India as per National Sample Survey Organization 2005 estimate 60 per cent farmers want to leave farming.
- This has direct impact on socio-economic society. Countries are facing social disturbance with growing local terrorism and riots for food. In India this is reflected by expansion of terrorism activities in various states. A Movement that encourages young individuals to crimes of extortion and dacoits which has cover 1/3rd of its total district during which they're present.
- But with available knowledge and experience, it's possible to form agriculture a sustainable livelihood means – but this may require intensive efforts at ground level – local level where agriculture exists and it's has to be made climate smart.

What CSA Means

Achievement of sustainable development goals and which integrates – social, economical and environmental development to satisfy challenge of providing sustainable

(a) Livelihood to farmers
(b) Food security to hungry millions
(c) Eradication of poverty.

It is composed of four pillars

- Sustainably increasing agriculture productivity and income
- Adapting and building resilience for climate change
- Removal of greenhouse gas emissions wherever possible
- It uses agriculture as a major tool for mitigation of Greenhouse Gases – CO_2 by laying emphasis on its unique capacity to absorb CO_2 and release Oxygen through photosynthesis process.

It reliable in achieving this through increased cropping, by reducing rain fed areas through integrated water and basin management and expansion of agriculture on wasteland, wetland, degraded fallow areas and introducing urban agriculture. It is a way for addressing the developmental efforts towards the technical, policy and investment condition related sustainable agricultural development to achieve food security under climate change along with eradication of poverty. But climate smart agriculture main focus is to act at local level where there is previously has impact of climate. Reasons need to act together at local level for climate smart agriculture

- ☆ Climate change has already adverse impact but there are available solutions – what's needed is to translate them at village level.
- ☆ There are many programmes like plans and missions but they need to happen at ground level and hence, despite of their existence need to start CSA at bottom level
- ☆ Farmers are intelligent and clever know what are their problems. Majority of farmers in India today are literate and capable of accepting new ideas and practices.

So, it is the local level (taluka level) development administration but they have to be prepared to develop their own production plan and prepare to believe in them that they are capable of implementing Climate Smart Agriculture. For environmental concern, global warming is increasing its emissions of GHG and non-willingness of lead players to provide curbs as agriculture offers a way to mitigation. Though photosynthesis is unique process, agriculture absorbs CO_2 from atmosphere and releases Oxygen. With increased productivity, area under agriculture and multiple cropping also increased, which reduce CO_2 and leads to sustainable livelihood and food security to hungry millions. Technologies, knowledge exists, and plans are prepared from top to middle level while vacuum at the bottom still remains unaddressed. Only some are benefited while the majority are left out of development process in spite of claiming "all inclusive" approach. This trend across the entire developing world is identical. Gaps exist in same village with same land and water resources between one farmer and another farmer. The progressive one makes profit, the average one fail, and some commits suicide or others migrate to urban centres – rest drag on. But at the end result overall productivity and income to farmers suffer.

Sustainable Livelihood

Livelihood means, provide enough to all to live or exist. It should not become less and there is gradual increase in income for better standard of life. Climate Smart and Sustainable Agriculture (CSA) is a means to provide this to farmers. CSA endures to ensure that despite change in climate and its adverse impact on crops/animals, income to farmers should not decrease. It provides opportunities to have multiple sources of income from agriculture and animal husbandry - mulch cattle and poultry, fisheries, when one fails, other supports. It provides opportunity to young members of family to acquire multiple skills, support for setting up microenterprise locally, based on demand and supply situation. It acts like safety net at the time of natural calamities by way of insurance for crops and animal husbandry along with employment in community projects.

Climate Smart Agriculture Involves

- Cropping pattern will be based on soil health and moisture analysis to support crops which can be sustained.
- Agro – advisory
- On predicted weather pattern long term, medium term and week to week basis given by Meteorology Department this prior to kharif and Rabi season
- After unexpected weather the changes have occurred for corrective action to prevent crop loss and livestock management.
- Crop production that contributes to food security by addressing current and projected climate change impacts through adaption and mitigation and provides an opportunity to win-win situation despite adverse changes.
- It provides institutional arrangement for mass communication and a way to bridge productivity gaps at local level between farmers by reaching out farmers at their door step.

History of CSA

- In 2009: The term Climate Smart Agriculture was developed.
- 2010: 1st Global conference on food security agriculture and climate in The Hague. The concept was presented.
- 2012: At the 2nd Global conference in Hanoi, Vietnam: Climate smart agriculture. Vulnerable people in developing countries.
- 2013: 3rd GC in Johannesburg, South Africa, discussion began on climate smart agriculture alliance.
- 2014: Climate summit in New York, the global alliance for Climate Smart Agriculture Action plan was presented.
- There have been 5 Climate smart and 88 agricultural global sciences conferences are held. Netherland-2011, Davis- 2013, France- 2015, Indonessia-2019.
- 2019: 5th Global science conference on CSA on transforming food system under climate change at Bali, Indonesia.

Impact of Global Warming

Our earth is heated by sunlight. Most of the sun's energy passes through the atmosphere, to warm the earth's surface, oceans and atmosphere. However, in order to keep the atmosphere's energy budget in balance, the warmed earth also emits heat energy back to space as infra-red radiation. A natural system known as the "greenhouse effect" regulates temperature on Earth. Just as glass in a greenhouse keeps heat in, our atmosphere traps the sun's heat near earth's surface, primarily through heat-trapping properties of certain "greenhouse gases". Over the past thousands of years the amount of greenhouse gases in our atmosphere has been relatively stable. A few centuries ago, their concentrations began to increase due to the growing demand for energy caused by industrialization and rising populations, and due to changing land use and human settlement patterns. The greenhouse effect refers to the change in the steady state temperature by the presence of gases that absorbs and emits infra-red radiation. The greenhouse gases trap heat within the troposphere. The gases are water vapour, carbon dioxide, ozone, methane, nitrous oxide and chlorofluorocarbons. Nitrogen, oxygen and argon make up 98 per cent of the Earth's atmosphere. But they do not absorb significant amounts of infra-red radiation and thus do not contribute to the greenhouse effect. Carbon dioxide (CO_2) constitutes about 72 per cent of total Greenhouses Gases and contributes the bulk of radioactive forcing.

Increase of burning of fossil fuel like coal, oil, gas and also wood *etc.* caused by industrial activities and deforestation have increased its concentration in the atmosphere. The concentration of carbon dioxide (CO_2) in the atmosphere has increased from 285 ppm at the end of the nineteenth century before the industrial revolution, to about 385 ppm in the new millennium. Methane is produced when vegetation is burned, digested or decayed with no oxygen present. Garbage dumps, rice paddies, and grazing cows and other livestock release methane. Nitrous oxide is released when chemical fertilizers are used in agriculture. Other gas is SF-6.

Most interesting part of these major emitters is CFC 12 ÷ 1 unit = 7000 CO_2), SF6 (1 SF = 23900 CO_2) are largely from industrialized nationals. GHG emission is also common through agriculture activities by the use of fertilizer, methane from cow dung or rice fields. But while calculating actual emissions from agriculture – no deduction are considered to the gases that each crop absorbs. Every crop, plant, tree or grass absorbs CO_2 and releases oxygen.

The Impact on India

Unpredictable Climate is a threat to sustainable development: Every single day, there is breaking news about natural calamities hitting some region in the world. This unpredictable climate is creating havoc around the world, destroying habitats and disturbing people's livelihood. Some of the recent natural disasters related to these phenomena in India are narrated below. It must be noted that our country is not new to droughts, cyclones *etc.*, but its frequency and intensity have increased abnormally in the new millennium. India like other countries in the world has its share of natural disasters.

Gujarat Earthquake, 2001

- ☆ The Kutch earthquake that shook Gujarat was one of the deadliest earthquakes to strike India. The region continues to simmer and has experienced several mild earthquakes and tremors since 2001.

Trail of Destruction

- ☆ The death toll: 19,727
- ☆ Injured: 166,000
- ☆ Homeless: 6 lakhs

- Houses destroyed: 3, 48,000
- Cattle killed: 20,000
- Estimated losses at: 1.3 billion
- Tsunami December 2004: affected the Andaman and Nicobar, Pondicherry, Kerala and Andhra Pradesh causing loss of agri-crops, cattle wealth, housing and livelihood.
- Mumbai Floods: 26th July 2005 – the city was paralyzed and floods in Maharashtra
- Surat Floods (2006): Estimated loss of Rs.22, 000 crore. City's infrastructure affected, high individual losses and crops like sugarcane (Rs. 4,000 crore)
- Heavy rains in 2007 in Rajasthan: with flooding and consequent breakout of diseases, loss of crops and cattle wealth.
- Bihar: 2008 – Koshi river overflow with dam in Nepal giving way and large areas of Bihar - UP got affected. Droughts 2009: Delayed monsoon caused drought in some states.
- 2010: Heavy floods in Northern India and un-seasonal showers and snow falling in some parts of India like Gujarat in 2010, the most important is that weather has become totally unpredictable.
- 2011 -13: Floods in Andhra Pradesh so and so far Odisha, land slide in Uttarakhand.
- 2014-15: The monsoon has got delayed.

Global Impact

The worldwide impact is equally grave.

- In the U.S. alone, nearly 1,000 tornadoes have killed many people and inflicted $9 billion in damage.
- The 2010 heat wave in Russia killed hundreds of people and led to a 40 per cent fall in the harvest of food grains.
- Floods in Australia and Pakistan killed thousands of people and devastated agricultural lands.
- Re-current droughts in China have eroded millions of acres of farmland.
- Tsunami in Japan – Nuclear plant affected and played havoc to the local habitat
- Recurrent and continuous famines in Ethiopia - Somalia and riots for food by hungry millions.
- Recurrent floods in South East Asia, Philippines, Indonesia, Thailand.
- Almost all nations small or big are affected, one way or another
- This is continuously happening with increasing intensity.

Such Impacts Create

- Severe famine or heavy floods

- Loss of life
- Loss of agriculture crops and animals

Effects of Climate Change

- Loss of livelihood
- Increased risk of diseases outbreak and germination of new viruses.
- Damage to infrastructure and communication particularly in rural areas.
- Setback to social and economic development and emergence of social turmoil with increased rural urban divide.
- Pushing farmers in rural areas again back below poverty line.

Current Indian Scenario

Dr A. K. Singh, Formerly, Deputy Director General of Indian Council of Agriculture Research has analyzed the overall Indian scene.

- No significant change in monsoon rainfall at All India level, some regional variations are noticed
- Increase in rainfall in west coast, north AP and NW India, decreasing trend in east M.P. and adjoining areas
- Increase in surface air temperature by 0.510 C during 1901-2007, accelerated warming during 1971-2009
- Mean temperature rise by 0.20 C per decade during 1971-2009, greater rise in minimum temperature than maximum
- One day extreme rainfall events are increasing
- Cyclonic storms in Bay of Bengal showed declining trend of 2 cyclones/decade during 1891-2008
- Deglaciation in the Himalayas
- Sea level rise in Indian Ocean 1.63 mm/year during 1993-2009 Indian Scenario: Future Projection
- Change in rainfall pattern by the end of the 21st century
- Increase in temperature by 2 to 4 0C
- Warming will be more pronounced over most of the land areas
- Maximum increase over northern India
- Relatively greater warming in winter and post monsoon seasons
- Frequency of cyclones during post monsoon seasons (2071 to 2100) is projected to much higher than baseline scenario (1961-1990)

Climate Smart Technologies

CSA includes a number of technological, policy, and institutional interventions (Aggarwal *et al.*, 2004) revolving around seed, water, energy, and nutrients and some risk-averting and risk-insuring instruments that increase the resilience and stability of agriculture and thus help farmers adapt to and reduce the risk of climate change. The selection of climate smart technologies for the present study was based on the principle of triple wins, as discussed above. This approach took the form of bio-intensive agriculture and sought to increase production per unit of land/water/nutrients. The end result is a reduction in the emission of greenhouse gases per ton of farm produce; sequestration of carbon dioxide in the soil; and increased crop resilience to heat, droughts, and floods. For the sake of clarity and better communication, we classify the various technologies and interventions into five groups: water-smart, energy-smart, nutrient-smart, crop-smart, and weather-smart interventions. The specifications of the technology basket are given follow.

1. Water-Smart Technologies

- ✰ Water-smart technologies
- ✰ Interventions that reduce water requirements to produce the same or a higher level of yield.
- ✰ Rainwater management
- ✰ In situ rainwater storage in rice paddies with 20–25 cm bunds. This technique is for rice only.
- ✰ Laser land levelling.
- ✰ System of rice intensification
- ✰ A 7- to 10-day-old seedling is transplanted at 20 cm spacing with 1–2 seedlings per hill.
- ✰ Furrow-irrigated raised bed
- ✰ Growing crops on ridges or beds. Irrigation is applied through furrows separating the beds.

A major impact of climate change on agriculture will be experienced in the form of water stress. Therefore, land development, water conservation and harvesting measures in both rain fed and irrigated agriculture, crop planting, and irrigating techniques, among others, could form a package of practices to help overcome the climate-imposed stress (Sharma *et al.*, 2006; Tyagi 2009). Specific technologies in this group include LLL, *in situ* rainwater management for rice (with 20–25 cm high bunds) and system rice intensification (SRI). The water-economizing and yield-maximizing potential of these technologies has been adequately established through experimental studies and on-farm trials in a number of locations in the IGP (Tyagi 2009), and farmers have acquired some familiarity with these technologies.

2. Energy-Smart Technologies

Energy-smart technologies -Technologies that help reduce energy consumption during land preparation without affecting yield levels. These also help reduce water requirements for crops. Direct-seeded rice -Dry seeds are sown either by broadcasting or drilling in line. Zero tillage/minimum tillage -The crop is seeded through seeders in an untilled field, and the crop residue is incorporated into the soil. At present, this technique is limited to wheat only.

Energy is the most important component for present-day agriculture. Although energy contributes to greenhouse gas emissions, the lack of it is a major barrier to agricultural development. Conservation agriculture is an energy-smart emerging intervention for sustainability (Lumpkin and Sayre 2009). The recent stagnation in productivity growth in the irrigated areas of the IGP has led to the development and refinement of resource-conserving technologies that save water and energy, improve soil health, and increase productivity. Practices to overcome energy constraints are zero tillage and direct seeding. Although these technologies focus on reducing energy use and the cost of production, the measures simultaneously save irrigation water by improving the soil moisture provided by rainwater, decreasing soil moisture evaporation, and decreasing pre-sowing irrigation. Each of these technologies has the potential to reduce water demand between 10 and 20 per cent and boost crop yields by a similar level. Incidentally, zero tillage, SRI and LLL were developed not only to save water but also to increase mechanization to cope with the growing demand for labor in agriculture.

3. Nutrient-Smart Technologies

Nutrient-smart technologies -Technologies those save/supplement/avoid chemical fertilizer use for crops and enrich carbon in the soil. Green manure -Cultivation of legumes in a cropping system. This practice improves nitrogen economy and soil health/quality. Integrated nutrient management -Integrated use of organic and chemical fertilizers to partially (25 per cent to 50 per cent) reduce NPK (nitrogen, phosphorus, and potassium) requirements without affecting productivity and improve soil health. Leaf colour chart -Standardized colour charts are used to identify nutrient deficiency to estimate fertilizer doses in different field locations.

Agricultural practices that ensure soil and nutrient management and carbon management along with nitrogen management can contribute to the conservation of resources, sequestration of carbon, and safeguarding of future food security (Lal *et al.,* 2011). There are a wide range of practices associated with this group of technologies, including green manure to supplement the applied chemicals and improve the physical condition of the soil, integrated nutrient management, and organic manure (farmyard manure and/or vermicompost) in a certain ratio (ICAR 2009). Some of these practices reduce emissions of greenhouse gasses and build organic carbon into the soil. The site-specific application of nutrients based on plant indicators such as leaf colour is a technology that has been standardized for various crops. Leaf colour charts are commercially available to help farmers monitor plant nutrient status and apply nutrients according to the plants' needs at different locations.

4. Stress Tolerant Seeds and Crop Mix-Smart Technologies

Introduction of stress tolerant crops and diversification -Tolerant crops withstand biotic and abiotic stresses and crop diversification reduces water demands and helps in harnessing nutrients from different soil layers. Drought-tolerant variety -Seed variety that is tolerant to drought or relatively dries weather conditions. Crop diversification (maize-wheat cropping) -Rice is replaced by maize on part of the land to economize on water use. Water- and nutrient-input-responsive cultivars were a major factor in triggering the Green Revolution in India, as they were the least expensive and highest yielding option.

Sustained efforts in recent years have led to the development of crop varieties that are resistant to biotic (pests and disease) and abiotic (flood and drought, salinity and heat) stresses for irrigated and rain fed conditions (ICAR 2009). Drought is a major climatic aberration that is experienced in both the states and therefore a technology representing drought-tolerant crop varieties was included in the assessment.

Further, it has been established that the current large-scale rice-wheat cropping system is creating long-term groundwater imbalances, requiring a switch to less water-dependent crops (Tyagi *et al.*, 2004; Ambast, Tyagi, and Raul 2006; Amarasinghe, Shah, and Singh 2007). Diversification of crops, cropping systems, and farming systems could be a potential response to overcome water scarcity and also to act as a mechanism to minimize weather-induced losses and stabilize incomes. Maize, with its high yield potential and low irrigation requirements, is considered a good option and has been included as a crop diversification measure.

5. Weather-Smart Instruments

Weather-smart instruments -Interventions that provide services related to financial security and weather advisories to farmers. Crop insurance Crop-specific insurance to compensate income loss due vagaries of weather. Weather advisories Information and communication technology based forecasting about the weather. The vagaries of weather events such as floods and drought are a major reason for fluctuating productivity and farm incomes. These events are more frequently experienced in Eastern IGP than in Western IGP. This situation is expected to worsen under the current climate-change scenario.

Risk management through the dissemination of weather advisories and climate information, together with weather-based insurance, is a non-structural intervention to reduce production losses and stabilize farmers' income. Weather forecasting and advisory services and their dissemination through information and communication technology have made it possible to offset farming losses. Therefore, weather advisories and crop insurance have also been included as possible instruments to stabilize incomes and improve farmers' capacity to adopt other productivity-enhancing technologies.

Other Technologies

Basic Element-Sensors

1. Sensor Networks

Sensors and their networks are in use in agriculture for crop monitoring, local weather monitoring, soil attributes monitoring, *etc.* Sensors are used mostly to measure air temperature, Humidity, ambient light, soil temperature, soil moisture, soil pH value, air speed, air direction, *etc.*

2. IoT

The Internet of Things, or IoT, is a system of interrelated computing devices, mechanical and digital machines, objects, animals or people that are provided with unique identifiers (UIDs) and the ability to transfer data over a network without requiring human-to-human or human-to-computer interaction. In IoT based smart farming, a system is built for monitoring the crop field with the help of sensors (light, humidity, temperature, soil moisture, *etc.*) and automating the irrigation system. The farmers can monitor the field conditions from anywhere.

3. Unmanned Aerial Vehicles

Drones can be put to excellent use in the agricultural industry. There are two types of drones namely Ground-based Aerial drones.

Applications of Technologies

1. Monitoring of Climatic Conditions

Most popular smart agriculture gadgets are weather stations, combining various smart farming sensors. Located across the field, they collect various data from the environment and send it to the cloud.

2. Crop Management

Sensors should be placed in the field to collect data specific to crop farming; from temperature and precipitation to leaf water potential and overall crop health, these can all be used to readily collect data and information for improved farming practices. Thus, you can monitor your crop growth and any anomalies to effectively prevent diseases or infestations that could harm your yield.

3. Cattle Management

Just like crop monitoring, there are sensors that can be attached to the animals on a farm to monitor their health and log performance. Smart agriculture sensors (collar tags) to deliver temperature, health, activity, and nutrition insights on each individual cow, as well as collective information about the herd.

4. Greenhouse Automation

A smart greenhouse can be designed with the help of IoT; this design intelligently monitors as well as controls the climate, eliminating the need for manual intervention. Smart sprinklers controller that allows you to manage your irrigation and lighting systems remotely.

5. Smartphone Tool

Smart phone such as a Camera which Provides pictures of leaf health, lighting brightness, chlorophyll measurement, and ripeness level. Also used for measuring Leaf Area Index (LAI) and measuring soil organic and carbon makeup. GPS which Provides location for crop mapping, disease/pest location alerts, solar radiation predictions, and fertilizing. Microphone helps with the predictive maintenance of machinery. Accelerometer is helps to determine Leaf Area Index. Also used as an equipment rollover alarm. Gyroscope which detect san equipment rollover.

6. Smart Apps

Plantix

Plantix is a mobile crop advisory app for farmers, extension workers and gardeners. It can diagnose plant diseases, pest damages and nutrient deficiencies affecting crops and offers corresponding treatment measures. Other apps such as crop doctor, My-rml, Honey bee.

Kisansuvidha

This app is Launched by Prime Minister Mr. Narendra Modi in 2016. It provides information on current weather and also the forecast for the next five days, market prices of commodities/crops in the nearest town knowledge on fertilizers, seeds and Machinery *etc.* The option to use the app in different languages makes it more widely accessible

CSA Application in India

1. Climate–Smart Villages (CSV)

It's the CGIAR Program on climate change, Agriculture and Food Security (CCAFS) CSV is a model of local action for climate risk management in farming communities that Promote adaption, Built resilience to climate stresses, Enhance food security. CSV's need a collaboration of Researcher, Local organisation, Policy makers and farmers. To select the most appropriate technologies and institutional intervention for Enhance productivity, Increase farmer's income, Achieve climate resilience and Ensure food security. The key focus of the CSV model is 1)To enhance climate literacy to of farmers and local stakeholders 2)Develop a climate resilient agricultural system by linking existing government to village development Schemes and investment 3)Promotion of combination of CSA practices and technologies. The four key component of CSV is Climate Information System, Local Knowledge and Institution, and Village Development Plans and Climate Smart Technology.

2. National Initiative on Climate Resilient Agriculture (NICRA)

NICRA is an ICAR project launched in 2011; under this project 151 villages are covered across the country. To take up long term strategic research to address the impact of projected climate change on Indian agriculture. Demonstrate the existing best practices to enable farmers to cope with current climate variability. First ever vulnerability atlas of India at district level which aid in prioritizing investment in vulnerable regions.

3. Initiatives of Krishi Vigyan Kendra's (KVK)

Initiatives are like V-KVK- which provides guidance to the farmers with help of mobile phone they can direct do talk with experts, SMS Advisory; forecasting the weather alert messages on the farmer's mobile phone and also Community radio, Weather Information and Hiring agriculture Implements.

4. Soil Health Card

Soil Health Card programme has been launched in 2015 by Honourable Prime minister Narendra Modi. The programme has been run by Department of Agriculture and Cooperation, SHC'S are provided by Indian Council of Research. Soil health card provides all the details of nutrient availability status of soil, no of plots, name of the borrower. It helps farmer and experts for crop and soil nutrient management, work as guide to the farmer.

Benefit of CSA

CSA helps to maintain stable yield with help of giving proper guidance to the farmer by providing information such as weather forecasting, market forecasting, SMS advisory system, adoption of new technologies to them. Drought buffering with help of Stress Tolerant Seeds and Crop Mix -Smart Technologies. Reduced field preparation cost by adopting laser levelling of land and zero tillage practices of land cultivation. Reduced soil erosion by adopting laser levelling of land and zero tillage practices of land cultivation. Climate Smart Agriculture is Climate change mitigation.

Disadvantages of CSA

Smart farming requires skills, and computer based intelligence, skills the average farmer would not necessarily have. Initial capital costs may be high and so it should be seen as a long-term investment. It may take several years before you have sufficient data to fully implement the system. Extremely demanding work particularly collecting and then analysing the data. Smart agriculture needs availability of internet continuously.

Conclusion

Even though CSA is difficult to adapt immediately require skill and knowledge, Cost requires several years for proper implementation. Climate change demands CSA to meet challenges in food security as well as the environmental issues. CSA is the way to curtail the effects of climate change, assured food security and sustainable and energy saving method.

REFERENCES

Alvin Chandra, Karen E. McNamara and Paul Dargusch (2018) Climate smart agriculture: perspectives and framings, Climate Policy, 18: 4, 526-541,

http: //www.fao.org/climate smart-agriculture/en/

https: //www.mouser.in/applications/smart-agriculturesensors/

https: //www.postscapes.com/smart-agriculture/

https: //www.iotforall.com/iot-applications-inagriculture/

https: //thingsboard.io/smart-farming/

Hochman, Z., Horan, H., Reddy, D.R., Sreenivas, G., Tallapragada, C., Adusumilli, R.; Gaydon, D.S., Laing, A., Kokic, P., Singh, K.K. *et al.*, Smallholder farmers managing climate risk in India: 2. Is it climate smart? Agric. Syst. 2017, 151, 61–72.CSA Sourcebook 2013 and 2016.

FAO Report 2015 and 2016.

Chapter 3

Climate Smart Agriculture: Innovative Extension Approaches, Strategies and Rural Advisory Services

M.S. Anarase, G.K. Sasane and P.A. Ghadage*

Department of Agricultural Extension and Communication, Post Graduate Institute, Mahatma Phule Krishi Vidyapeeth, Rahuri – 413 722, Maharashtra

Climate change is the most serious environmental issue to deal against hunger, malnutrition, disease and poverty in global level. Agriculture is the most vulnerable and sensitive sector affected by climate change because of its dependency on climatic parameters like rainfall and temperature *etc.* To alleviate the challenges posed by climate change, agriculture has to become "climate smart", that is, sustainably increase agricultural productivity and incomes, adapt and build resilience against climate change, and reduction or removal of greenhouse gases emissions, wherever possible. Despite the recognised importance of Climate Smart Agriculture (CSA), the dissemination and uptake of climate smart technologies, tools and practices is still largely an ongoing and challenging process. The adaption and dissemination of climate-related knowledge, technologies and practices to local conditions, promoting joint learning by farmers, researchers, extension worker and widely disseminating CSA practices, is much crucial and critical. In this regard, Chapter is mainly focusing on Climate change scenario in Indian and World agriculture context and their impact on agriculture sector. Climate Smart Agriculture with respect to Extension approaches used for CSA, Extension strategies for transfer climate related knowledge and Extension methods for learning and development of capacity. Different Project include National Innovation on Climate Change and Climate Change Adaptation. Different Innovative Extension approaches and Rural advisory services used in Climate Smart Agriculture (CSA).

Keywords: *Climate smart agriculture (CSA), Climate change, Extension strategies, Approaches and Rural advisory services.*

Introduction

Agriculture in developing countries must undergo significant transformation if it's to fulfil the growing and interconnected challenges of food insecurity and temperature change (FAO, 2010). Global climate change is that the most severe challenge that affects development in 21st century. It's one in every of the main threats to humankind and affects many sectors like forestry, agriculture, environment and human lives similarly. Global climate change has caused severe and possibly permanent alterations to our planet's geological, biological and ecological systems. Agriculture is additionally vulnerable to the global climate change which directly and indirectly affecting the farming practices and crop yields. Farmers face many problems because of global climate change. The croplands, pastures and forests that occupy approximately 60 per cent of the earth's surface are progressively being exposed to threats from increased climatic variability. As climatic patterns change, there comes change within the distribution of agro-ecological zones, habitats, distribution patterns of plant diseases and pests, fish populations and ocean circulation patterns which may have significant impact on agriculture and food production. The challenge of rapidly boosting productivity is compounded by the present and expected impacts of temperature change. Changes to precipitation and temperature, especially in marginal areas, are expected to scale back productivity and make production more erratic (Cline,2008). Ensuring that agriculture becomes climate smart could be a priority for addressing the requirement for adequate, nutritionally balanced food for a growing and more demanding population during a situation of resource limitations, and temperature change and variability. Consequently, there's a requirement to simultaneously improve agricultural productivity and reduce yield variability over time under adverse atmospheric condition. A proposed means to realize this can be increased adoption of a 'Climate Smart Agriculture' (CSA) approach (FAO, 2010). CSA, which is defined by its intended outcomes, instead of specific farming practices, consists of three main pillars: sustainably increasing agricultural productivity and incomes; adapting and building resilience to global climate change and reducing and/or removing gas emissions relative to traditional practices (FAO, 2013a). So, CSA is a crucial approach in agriculture to cater to the foremost challenging issue of the globe. Much research has been conducted on the biophysical aspect of global climate change but socioeconomic research regarding the impact of those CSA practices is especially lacking. So, there is a necessity of strong extension network in climate smart agriculture to vary the behaviour of farmers or to produce them different location specific adaptation and mitigation strategies.

Climate Change

What is Climate Change?

"A change of climate which is attributed directly or indirectly to human activity that alters the composition of the global atmosphere and which is in addition to natural climate variability observed over comparable time periods." (United Nations Framework Convention on Climate Change -UNFCCC 1992).

Causes of Climate Change

Table 3.1: Causes of Climate Change

1) Natural Causes	2) **Anthropogenic Causes**
1) Continental drift	**1) Greenhouses Gases:**
2) Volcanoes	1) Carbon dioxide (CO_2)
3) The Earth's Tilts	2) Methane (CH_4)
4) Ocean Currents	3) Nitrous oxide (NO_2)
5) Intensity of Solar Radiation	4) Chloro floro carbons (CFCs)
	5) Ozone (O_3)
	6) Water Vapours (H_2O)
	2) Land Use Change:
	Deforestation and Urbanization

Climate Change and Agriculture Sector

Agriculture is that the backbone of financial system in most of the countries. Additionally, to food and material, agriculture also provides employment opportunities to large population. Climate change directly affects agricultural production as this sector is inherently sensitive to weather conditions and is one amongst the foremost vulnerable sectors at the danger and impact of world climate change (Parry *et al.*, 2005). Agricultural production has always been closely linked with variations in weather. Climate change is projected to own significant impacts on conditions affecting agriculture, including temperature, greenhouse emission, glacial run-off, precipitation and interaction of those elements. These conditions determine the carrying capacity of the biosphere to supply enough food for the human population and domesticated animals. the general effect of temperature change on agriculture depends on the varied measures adopted to balance these effects.

Generally, climate change could affect agriculture in several ways: productivity, in terms of quantity and quality of crops; agricultural practices, through changes of irrigation and agricultural inputs like herbicides, insecticides and fertilizers; environmental effects, particularly in relation of frequency and intensity of soil drainage (leading to nitrogen leaching), wearing, reduction of crop diversity; rural space, through the loss and gain of cultivated lands, land speculation, land renunciation, and hydraulic amenities; and adaptation, (organisms may become more or less competitive, like flood-resistant or salt-resistant styles of rice). In general, temperature change could affect agriculture in several ways:

- Productivity, in terms of quantity and quality of crops
- Agricultural practices, through change of water use (irrigation) and agricultural inputs like herbicides, insecticides and fertilizers

- Environmental effects, specifically, in relation of frequency and intensity of soil drainage (leading to nitrogen leaching), wearing, reduction of crop diversity
- Rural space, through the loss and gain of cultivated lands, land speculation, land renunciation, and hydraulic amenities.
- Adaptation, organisms may become more or less competitive, likewise as humans may develop urgency to develop more competitive organisms, like flood-resistant or salt-resistant forms of rice.

The Inter-Governmental Panel on temperature change (IPCC), in its third assessment report published in 2001, concluded that the poorest countries would be hardest hit, with reductions in crop yields in most tropical and sub-tropical regions because of decreased water availability, and new or changed insect pest incidence (IPCC, 2001). In Africa and geographical region many rainfed crops are near their maximum temperature tolerance, so their yields are likely to fall sharply for even small climate changes. Decreases in agricultural productivity of up to 30 per cent over the 21st century has been projected. Marine life and also the fishing industry also will be severely affected in some places. In Asia, agricultural crop yield is predicted to say no by 5-30 per cent by 2050s because of rising temperature in Himalayas and this decline in agricultural yield will result in food insecurity, which becomes a significant future problem for kinsmen (IPCC, 2007).

Impact of Climate Change on World Agriculture Sector

Global climate change has direct impact on food production across the world. Increase in mean seasonal temperature can reduce the duration of the many crops and hence reduce final yield. In areas where temperatures are already near the physiological maxima for crops, warming will impact yields more immediately (IPCC, 2007). Developing countries, many of which have a median temperature that's already near or above crop tolerance levels, are predicted to suffer a mean 10 to 25 per cent decline in 3 agricultural productivity by 2080s. Rich countries, which have typically lower average temperatures, will experience a way milder or perhaps positive average effect, starting from 8 per cent increase in productivity to six decline. Individual developing countries face even larger declines. India, as an example, could see a drop of 30 to 40 per cent (Mahato, 2014). global climate change may cause a rise in both crop and livestock productivity in mid- to high latitudes (IPCC, 2007) and a decrease in tropical and subtropical areas. the foremost affected areas from temperature change are economically vulnerable countries, already food insecure and a few important food exporting countries. this can induce significant changes in trade, impacting prices and therefore the situation of net food importing countries. Consequently, global climate change is predicted to extend the gap between developed and developing countries as a result of more severe impacts in already vulnerable developing regions, exacerbated by their relatively lower technical and economical capacity to reply to new threats. Smallholders and pastoralists will suffer complex, localized impacts (IPCC, 2007). International Food Policy Research Institute (IFPRI), it'll cause a rise of between 8.5 and 10.3 per cent within the number of malnourished children all told developing countries, relative to scenarios without temperature change (Nelson *et al.*, 2010).

Agriculture Sector and Climate Change in India

India is an agriculture dependent country and quite two-third of its population depends on agriculture for his or her survival. Agriculture contributes to approximately 14 per cent to India's GDP. India may be a large country with a various climate. Diverse seasons mean diverse crops and farming systems. there's a high dependency of agriculture on the monsoon rains and a detailed link exists between climate and water resources. The impacts of temperature change are global, but countries like India are more vulnerable see able of the high population reckoning on agriculture. In India, significant negative impacts are implied with medium-term (2010-2039) global climate change, predicted to cut back yields by 4.5-9 per cent, betting on the magnitude and distribution of warming. Since agriculture makes up roughly 16 per cent of India's GDP, a 4.5-9 per cent negative impact on production implies a value of global climate change to be roughly up to 1.5 per cent of GDP each year.

People and their livelihood are directly or indirectly full of global climate change. Global climate change poses a right away and growing threat to the livelihoods of a lot of people in India. Poor rural households, whose livelihoods depend predominantly on agriculture and natural resources, will bear a disproportionate burden of adverse impacts of global climate change (Kates, 2000; Mendelsohn *et al.*, 2007; Satapathy *et al.*, 2011). Yield reductions are predicted in wheat and rice because of temperature rise in key growing regions. 2010 was the warmest year on record in India (+0.93°C) since 1901. The productivity of most cereals would decrease thanks to increase in temperature and CO_2 and also the decrease in water availability. there'll be a projected loss of 10-40 per cent in crop production by 2100 if no adaptation measures are taken.

Climate Smart Agriculture

What is Climate Smart Agriculture

Climate smart agriculture (CSA) may be defined as an approach for transforming and reorienting agricultural development under the new realities of climate change (Lipper *et al.*, 2014). The Foremost commonly used definition is provided by the Food and Agricultural Organisation of the United Nations (FAO, 2013a), which defines CSA as "agriculture that sustainably increases productivity, enhances resilience (adaptation), reduces/removes GHGs (mitigation) where possible, and enhances achievement of national food security and development goals".In this definition, the principal goal of CSA is identified as food security and development (FAO 2013a, Lipper *et al.*2014); while productivity, adaptation, and mitigation are identified because these three interlinked pillars necessary for achieving this goal.

History of Climate Smart Agriculture

- ☆ 2009: Term Climate Smart Agricultural development

- 2010: 1st Global Conference on Food Security, Agriculture and Climate Change in The Hague - the concept of CSA was presented.
- 2012: At the 2nd Global Conference in Hanoi, Vietnam: Climate Smart Agriculture Sourcebook advanced the CSA concept desiring to benefit primarily smallholder farmers and vulnerable people in developing countries.
- 2013: 3rd Global Conference in Johannesburg, South Africa, discussions began on a climate smart agriculture alliance.
- 2014: Climate Summit in New York, the Global Alliance for Climate Smart Agriculture Action plan was presented

There have been three Climate Smart Agricultural Global Conferences:

- Wageningen, Netherlands, Oct 24-26 2011
- Davis, CA March 20-22 2013
- A third in LeCorum Montpellier France, March 16-18. 2015

Concept of Climate Smart Agriculture

- An integrated approach to developing technical, policy and investment conditions to achieve sustainable agricultural development for food security under climate change.
- It integrates the three dimensions of sustainable development (economic, social and environmental) by jointly addressing food security and climate challenges.
- CSA brings together practices, policies and institutions that are not necessarily new but are used in the context of climatic changes, which are unfamiliar to farmers.
- What is also new is the fact that the multiple challenges faced by agriculture and food systems are addressed simultaneously and holistically, which helps avoid counterproductive policies, legislation or financing.

Why Climate Smart Agriculture Needed?

Climate smart agriculture helps within the following ways.

- Sustainably increase agricultural productivity and incomes;
- Adapt and build resilience to climate change
- Reduce and/or remove greenhouse gas emissions, where possible.

Climate smart agriculture is not a new agricultural system, nor a group of practices. It's an innovative approach for charting development pathways which will make the agriculture sectors more productive and sustainable and better able to contribute to climate change adaptation and mitigation.

Objective of Climate Smart Agriculture

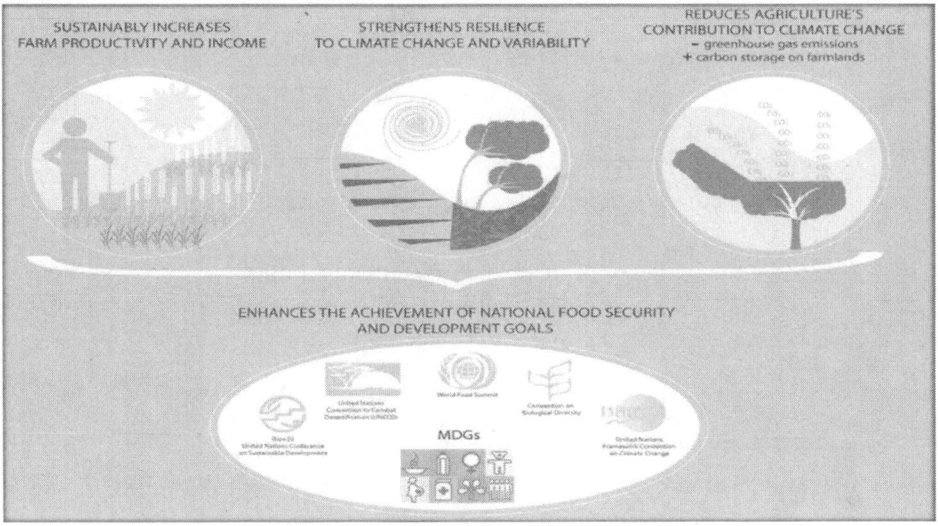

Three Pillars of Climate Smart Agriculture

1. Productivity

CSA aims to sustainably increase agricultural productivity and incomes from crops, livestock and fish, without having a negative impact on the environment. This, in turn, will raise food and nutritional security. A key concept related to raising productivity is sustainable intensification.

2. Adaptation

CSA aims to reduce the exposure of farmers to short-term risks, while also strengthening their resilience by building their capacity to adapt and prosper in the face of shocks and longer-term stresses. Particular attention is given to protecting the ecosystem services which ecosystems provide to farmers and others. These services are essential for maintaining productivity and our ability to adapt to climate changes.

3. Mitigation

Wherever and whenever possible, CSA should help to reduce and/or remove greenhouse gas (GHG) emissions. This implies that we reduce emissions for each calorie or kilo of food, fibre and fuel that we produce. That we avoid deforestation from agriculture. And that we manage soils and trees in ways that maximizes their potential to acts as carbon sinks and absorb CO_2 from the atmosphere.

Overview of Climate Smart Agriculture

- ✰ Addresses the complex interrelated challenges of food security, climate change, and identifies integrated options that make synergies and reduce trade-offs

- ☆ Recognizes that these options are going to be shaped by specific country contexts and capacities still as socio- economic and environmental situations.
- ☆ Assesses the interactions between sectors and the therefore the needs of various stakeholders.
- ☆ Identifies barriers to adoption (esp. for farmers), and provides appropriate solutions in terms of policies, strategies, actions and incentives
- ☆ Seeks to create enabling environments through a stronger alignment of policies, investments and institutions.
- ☆ Strives to realize multiple objectives with the understanding that priorities must to be set and collective decisions made on different benefits and trade-offs
- ☆ Prioritizes the strengthening of livelihoods (esp. those of smallholders) by improving access to services, knowledge, resources (including genetic resources), financial products and markets.
- ☆ Addresses adaptation and builds resilience to shocks, especially those related to climate change.
- ☆ Considers climate change mitigation as a potential secondary co-benefit, especially in low-income, agricultural-based populations.
- ☆ Seeks to spot opportunities to access climate-related financing and integrate it with traditional sources of agricultural investment finance.

Climate Smart Agriculture and Rural Advisory Services

Agriculture, as both a section of human action in danger from Climate change likewise as a driver of climate and environmental change, features prominently within the climate change scenario. To alleviate a number of the complex challenges posed by global climate change, agriculture (including forestry and fisheries) must become "climate smart", that is, sustainably increase agricultural productivity and incomes, adapt and build resilience to global climate change, and reduce and/or remove greenhouse gases emissions, where possible (FAO, 2013a). Climate smart agriculture is defined as an approach for transforming and reorienting agricultural development under the new realities of global climate change (Lipper *et al.*, 2014). it's defined as "agriculture that sustainably increases productivity, enhances resilience (adaptation), reduces Greenhouse Gases (GHGs) where possible and enhances achievement of national food security and development goals" (FAO, 2013b). Here productivity, adaptation and mitigation are three interrelating pillars in achieving goal of food security and development in CSA.

CSA is an integrative approach to deal with these interlinked challenges of food security and variations in climatic parameters that explicitly aims for 3 objectives:

1. Sustainably increasing agricultural productivity, to support equitable increases in farm incomes, food security and development;
2. Adapting and building resilience of agricultural and food security systems to global climate change at multiple levels;

3. Reducing greenhouse emission emissions from agriculture (including crops, livestock and fisheries).

Rural Advisory Services (RAS) contribute to achieving climate smart agriculture (CSA) by disseminating climate information and technologies and knowledge on production practices for climate adaption through innovative approaches, like plant clinics and participatory video and Digital Green. Extension system play a formidable role in supporting CSA through the technology development, knowledge dissemination, strengthening farmers' capacity, facilitation and brokering, and advocacy and policy support.

1. Increasing productivity sustainably and improving adaptation through the advancement of technology and dissemination of information.
2. Building resilience by building human and social resources for farmers and providing support services.
3. Supporting the adaptation and mitigation of climate change by facilitation.
4. Monitoring, advocacy and policy support.

Innovative Extension Approaches used in Climate Smart Agriculture (CSA)

Climate Smart Agriculture (CSA), is an innovative approach for address the urgent needs of mitigation of climate change adaptation, building resilience, and food security. But there is kind of lacuna is a lack of location-specific tools, long-term experiences and a favourable enabling environment are barriers to the effective implementation of CSA, there are a number of climate smart technologies and practices that are known and available. CSA considerations in extension strategies can still be considered as new. Addressing these global challenges requires strong generation, adaptation and use of new knowledge, which involves interaction and support from a wide range of organisations.

These new challenges also mean that extension systems need to tackle a variety of objectives that include, but go well beyond, transferring new technology. This encompasses the need to link more effectively and responsively to domestic and international markets (food, feed, fibres, *etc.* and/or carbon); reduce the vulnerability and enhance the voice of the rural poor; promote environmental conservation; build linkages between farmers and other agencies; and institutional and organisational development to support the bargaining position of farmers by, for example, forming farmer groups. Today, extension comes "in many sizes and shapes" and a distinction between the extension approaches as such (*e.g.*, participatory training approach, However, all extension systems share the common challenge of how to best respond to climate change.

There are several paths through extension systems can contribute to CSA. In addition, reach and impact potential, two negatively correlated indicators, are of primary importance and differ between extension approaches, *i.e.*, generally, the higher the reach, the smaller isthe impact and vice versa. Mass media often suits simpler messages while intensive interactions through farmer field schools can be more effective for complex knowledge. Choice of approach combinations can influence the ability of extension services to contribute to food security and income, adaptation and resilience, and climate change mitigation.

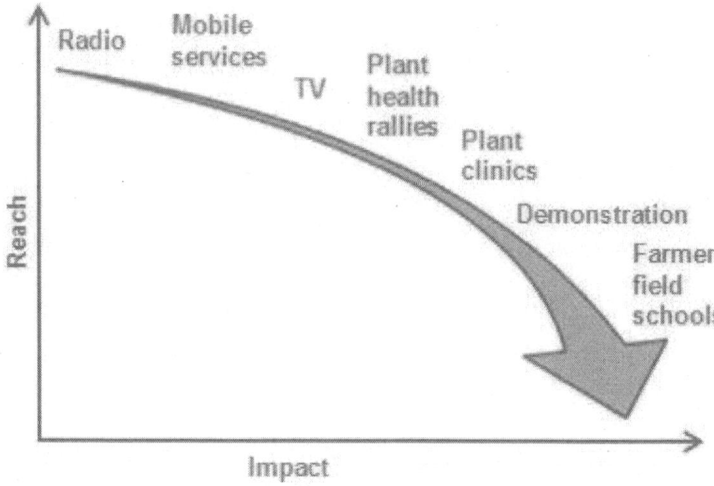

Figure 3.1: Impact of Extension Approaches (Sala *et al.*, 2016).

Table 3.2: Different Approaches Used in Extension for Climate Smart Agriculture

Sl.No.	*Approaches*
1.	Climate Awareness programmes/Campaigns, Exhibitions
2.	Climate Trainings
3.	Climate workshops
4.	Plant health Rallies
5.	Climate Farmers field schools (FFS)
6.	Field visits to progressive farmers
7.	Demonstration on different adaptation or mitigation practices
8.	Dissemination of appropriate climate resilient technology (such as portable soil testing kits, farm mechanization equipment for small holdings, grain storage bags, improved crop varieties *etc.*), irrigation management.
9.	ICT-supported network
10.	Participatory crop planning
11.	Appointment of climate manager at the village level
12.	Appointment of monsoon manager at the district level
13.	Use of indigenous technical knowledge (ITKs)
14.	Establishment of plant clinics
15.	Climate smart villages

Climate Awareness Mass Media Campaigns

Creating the awareness through mass media campaign is able to high reach of masses such approach is very famous and attractive for creating awareness on climate change. But we know, there is lack of field extension personnel in national system of extension and limited resources to reach extensive numbers of farmers spread widely across geographical locations. To overcome these constraints, extension can be more efficiently performed using mass media. Extension with mass media useful for creating awareness or deliver the information in simple way. A participatory mass extension campaign in Malawi, for example, has shown positive impact of radio messages on farmers' understanding of climate change and its effects, on how to produce high quality compost or about the importance of compost for building resilience (Mloza Banda, 2014). Such type of example showing that extension approaches that are best suited for raising the awareness and the potentially to contribute for climate mitigation, adaptation and increasing the food security.

Climate Training

Training related to the climate is important to update their knowledge related to climate change, their impacts, and consequences or on different adaptation and mitigation strategies. Climate smart agriculture is a new and Innovative approach so extension service providers should be trained on this aspect.

Plant Clinics

The plant clinic is an innovative approach is similar to human health clinics through which extension system allow direct exchange information between extension workers and farmers on "any problem and any crop". Plant clinics are a medium for facilitating face to-face exchange and two-way flow of knowledge and information between extension workers and farmers and link to other components of a plant health system (Boa *et al.*, 2015). The various crop problems like pest and diseases brought to plant clinics. Plant doctors should be knowledgeable about farmers and farming conditions, speak the local language, know what exact solution and inputs are available. Plant doctors are usually trained extension workers and remain employed by their current organisations.

Climate Farmers Field Schools (FFS)

The Farmer Field School (FFS) is a participatory, non-formal extension approach based on experiential learning that puts farmers and their demands at the centre (FAO, 2002). It provides farmers with a low-risk setting to experiment with new agricultural management practices, discuss and learn from their observations, which allows them to develop new practical knowledge and skills, and improve their individual and collective decision-making (Settle *et al.*, 2014). FFSs have also integrated elements of climate change adaption, such as the FAO FFS program on Integrated Plant and Pest Management (IPPM) that promoted improved and adapted varieties and agro-forestry practices in Mali and Niger (FAO, 2015). Climate Field Schools in Indonesia raised awareness of climate change and promoted solutions to cope with changing rainfall patterns, such as recording and interpretation of on-farm rainfall measurements and in-field water harvesting (Winarto *et al.*, 2008).

Plant Health Rally Approach

The plant health rally approach, first described by (Bentley *et al.*, 2003), is complementary to the plant clinic approach as it differs in terms of reach, impact and complexity of the messages that it can transmit. It is an extension method for quickly raising awareness about major agricultural danger or threats on important crops, to promote the use of improved agricultural practices, and to collect feedback from farmers on major issues which affect production. Plant health rallies are run by local extension workers and it is held in public spaces and are open to everybody.

Contingency Crop Planning

Contingency Crop Planning is form of calendar is very useful in terms of crop planting, scheduling of irrigation and plant protection measures for farmers. It is the important document that includes the key aspects of crop management and cultural practices. It is predicting on the bases of local weather conditions for the local crops grown in that particular area/location. It is prepared with the help of agro-meteorological department.

ICT Supported Network

The term Information and Communication Technologies (ICTs) playing acrucial role as a medium for disseminating information and communication on climate change awareness, adaptation and mitigation measures. But, the availability, accessibility and adoption of ICTs is varied between areas, developed and developing countries, urban and rural areas and within rural areas themselves. Farmers' ability to perceive climate change is a key precondition for their choice to adapt (Gbetibouo, 2008). Application of mobile phones, videos, radios *etc.* is helpful to address the issue of climate change by creating awareness among the farmers about the availability of different adaptation and mitigation strategies.

Farmer-to-Farmer Extension (F2FE)

Farmer-to-Farmer Extension (F2FE) is the provision of training by farmers to farmers, often through the creation of a structure of farmer-promoters and farmer-trainers (Scarborough *et al.*, 1997). F2FE offers great promise for effectively scaling up climate smart agriculture (CSA). F2FE is helpful to empowers farmers as agents of change and it is helps to increase the level of adoption because farmers are more willing to learn from their colleagues than from extension personnel (Franzel *et al.*, 2015). F_2FE programmes contribute to all CSA, that is, they help improve productivity, build resilience and reduce greenhouse gas emissions.

Climate Smart Villages (CSVs)

CSVs are the developed villages or models of local actions that ensure food security, promote adaptation and build resilience to climatic stresses. CSV have four components: climate information services; local knowledge and institutions;

Key Interventions in a Climate Smart Village

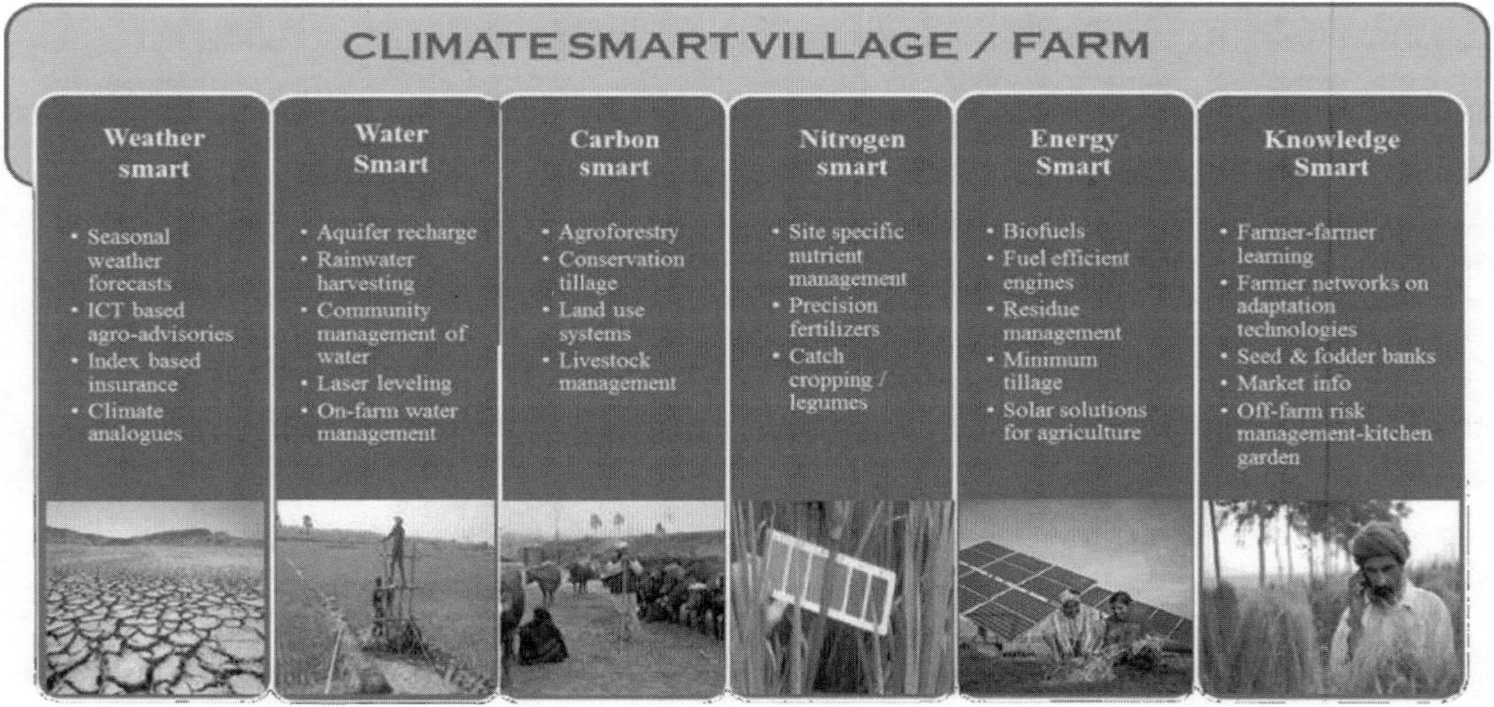

village development plans; and climate smart technology. The location of a CSV is selected based on its climate risk profile and the willingness of farmers and local governments to participate in the project.

Climate Smart Agriculture in India

Climate Smart Agriculture is not only an option but also necessity for making farmers more and more "climate smart". In the line with world, the Indian government also took some initiatives to deal with the several challenging issue in the context of climate change scenario. National Action Plan on Climate Change (NAPCC) was launched in 2008 with eight missions. Among them the National Mission for Sustainable Agriculture (NMSA) to make agriculture productive, sustainable and remunerative and climate resilient and to develop capacity of farmers and stakeholders in the domain of climate-change adaptation and mitigation measures. National Innovations on Climate Smart Agriculture was started by ICAR to enhance the resilience of Indian agriculture to climate change with specific objectives and components in 2011 in 100 selected most vulnerable districts. The CGIAR Research Program on Climate Change, Agriculture and Food Security (CCAFS) are leading projects in collaboration with national agriculture research and extension systems. They initiated Climate Smart Village (CSVs) programmes in Haryana, Punjab, Bihar, Karnataka and Odisha in 2011. In Haryana, CSVs were implemented in close collaboration with NICRA and through a range of innovative partnerships in 2011. The International Maize and Wheat Improvement Centre (CIMMYT) and CGIAR centres, farmer cooperatives, the Haryana Department of Agriculture and policy planners have piloted several CSV in Haryana as research and learning sites. Sustainable intensification and conservation agriculture-based management systems are the key areas in which CIMMYT and our partners work together. The CCAFS project is developing customised, region-specific crop insurance schemes based on short-term and long-term climatic risks.

The Government of India launched schemes for farmers like soil health card in 2015. Under the scheme, the government plans to issue soil cards to farmers which will carry crop-wise recommendations of nutrients and fertilisers required for the individual farms to help farmers to improve productivity through judicious use of inputs. ICT tools also used in climate smart agriculture, such as the use of mobile phone, portals, radio *etc.* The e-Arik project was initiated in 2007 in North East India aiming to disseminate climate smart agricultural practices and also to achieve food security.

Extension Strategies for Transfer of Climate Related Knowledge

Effective adaptation of the CSA advisories generated by different project can only be achieved if farmers had sufficient awareness and knowledge on climate change issues, like what is climate change, how it is affecting the agriculture, what are the consequences, impact and so on. So, the different strategies, projects and methods are used some of them are discussed below.

1. SMS or Short Messaging Services

Short text messages of 160-164 characters in the local language (Marathi) is sent to registered mobile numbers of farmers. Maximum two SMS per week based on weather advisories and contingency plans sent to the farmers.

2. Climate Wallpapers

One-page advisories in the form of tables or posters related to weather prediction of agricultural operations needed to be performed were pasted on common display boards of villages to provide advisory services to the farmers. In CCA project, these advisories were named Krishi Salah.

3. Climate Voice Messages

SMS is converted into voice messages in the areas with low literacy rate and disseminated to the farmers.

4. Use of Public Addressing System (PAS)

It is a good tool ford is semination of any urgent information in villager's area where SMS was not delivered due to network and electricity problems, farmers were informed about the critical climatic conditions or agro-advisory services is given with the help of this. For example, if rainfall is expected the next day, farmers would be immediately informed so that they could plan their farming activities accordingly. In Pathar, one cluster of Sangamner block, one permanent PAS is located in a temple of the village for public announcements related to climatic conditions.

5. Climate Group Meetings

Farmers organised in a group because it is easy to connect them and helpful for to dissemination of climate-related information in a group.

6. Exposure Visits

Organization of exposure visits for the farmers in the fields of progressive farmers such as those who scientifically cultivate the crop, use organic slurry for their crop, use drip or sprinkle irrigation *etc.* so that also motivated to adapt climate change.

7. Climate Workshops

Climate related workshop certain topics like preparation of water ponds, pomegranate cultivation, use of organic fertiliser, custom hiring centre *etc.* should be conducted to make farmers aware.

Description of Various Projects

1. Climate Change Adaptation (CCA) Project

This project was launched by Watershed Organisation Trust (WOTR) NGO in 2009 to dealing the consequence of changing scenario climate change on agriculture sector because agriculture is one of most sensitive sectors affected by climate change. Most of the India's population depends on agriculture directly or indirectly for survival. Variations of temperature and rainfall affect agriculture, which is directly linked to the Indian economy. So, it is necessary to deal with the changing scenario of climate change in agriculture. CCA was launched in the semi-arid and rainfed regions of Maharashtra, Telangana, Andhra Pradesh and Madhya Pradesh. There are different partners across sectors and scales-Swiss Agency for Development and Cooperation (SDC), the National Bank for Agriculture and Rural Development (NABARD), respective state governments as financiers and enablers; the Indian Meteorological Department (IMD), The Central Research Institute for Dryland Agriculture (CRIDA), the World Agro-Forestry council (ICRAF) and the Mahatma Phule Agriculture University (MPKV) and Bharati Vidyapeeth Institute of Environment Education and Research (BVIEER) as technology and knowledge partners, and WOTR which implements it generates and validates knowledge for use. The project was closed in 2015.

2. National Initiative on Climate-Resilient Agriculture (NICRA)

National Initiative on Climate Resilient Agriculture (NICRA) was launched during February 2011 by Indian Council of Agricultural Research (ICAR) with funding from Ministry of Agriculture, Government of India. This mega project has three major objectives: strategic research, technology demonstrations and capacity building. The major aim of NICRA is to enhance the resilience of Indian agriculture, covering crops, livestock and fisheries to climatic variability and climate change through development and application of improved production and risk management technologies; to demonstrate the site-specific technology packages on farmers' fields for adapting to current climate risks; and to enhance the capacity of scientists and other stakeholders in climate-resilient agricultural research and its application.

Objectives

- ☆ To enhance the resilience of Indian agriculture covering crops, livestock and fisheries to climatic variability and climate change through development and application of improved production and risk management technologies.
- ☆ To demonstrate site-specific technology packages on farmers' fields for adapting to current climate risks.
- ☆ To enhance the capacity of scientists and other stakeholders in climate-resilient agricultural research and its application.

The Project Comprised of Four Components

- ☆ Strategic research through network as well as Sponsored/Competitive Grants mode
- ☆ Technology demonstrations on farmers' fields to cope with current climate variability
- ☆ Knowledge management

- ✰ Capacity building of different stakeholders Climate Change Knowledge Network in Indian Agriculture (CCKN-IA) CCKN was German development cooperation (GIZ) and the Ministry of Agriculture and Farmers Welfare project established in 2013 in selected districts of three states – Maharashtra, Jharkhand and Odisha – to establish ICT-enabled approaches for knowledge exchange on climate-change adaptation in Indian agriculture. It promotes interactive dialogues among different stakeholders to establish and enhance linkage between climate change and sustainable development in agriculture.

Extension Methods for Learning

Climate Field Group Visits

Climate Field Group Visits is helpful to collect information about actual field condition during field visits, different agro-meteorological issues, ongoing government schemes *etc.*

Farmer Interest Groups (FIGs)

Farmer Interest Group (FIG) is a self-managed, independent group of farmers with a shared goal and interest. The members work together to achieve this goal by pooling their existing resources, gaining better access to other resources and to share in the resulting benefits. There are variety groups in the villages area *i.e.*, men and women such as the group of poultry-rearing farmers, goat-raring farmers, women SHGs, *etc.* FIGs make it easy to farmers as well as service providers.

Climate Trainings

Climate training is important for the agricultural extension workers to decide how to use NICE platform to give advisory services to the farmers and how to collect feedback of the farmers in CCKN project. Rain gauges installed in farmers' field and as per the norm of Indian Meteorological Department (IMD), KVK experts are in direct contact with these farmers to collect rainfall data. IMD provided technical material and field training to KVK experts and farmers on importance of rain gauges and how to install rain gauges at field level. Training is helpful to the villager's youths or farmers of the village how to read data from automated weather station or how to read temperature or how predict that information.

Informative Crop Calendar

It is a type of calendar which provide all agricultural practices package (information) of the crops. It includes planning, irrigation scheduling and plant protection measures based on the weather-based management. Example under CCA project, a crop calendar prepared with the help of CRIDA for the key crops grown on the Akole or Sangamner block of Ahmednagar. It includes weather-specific crop advisory services of specific crops like for paddy, finger millet, groundnut *etc.*

Block Contingency Planning

Contingency plans are technical documents containing integrated information on agriculture and allied sectors *i.e.*, horticulture, livestock, poultry, fisheries and technological solutions for all the major weather-related aberrations including extreme events *viz.*, droughts, floods, heat wave, cold wave, untimely and high-intensity rainfall, frost, hailstorms, pest and disease outbreaks. It is helpful to the effectively tackling local impacts is the key motivation for this entire innovative approach.

Information and Communication Tools

Information and Communication Tools like the NICE (Network for Information on Climate Exchange) platform is an IT-based system that facilitates collecting and disseminating up-to-date and relevant information to the farmers for take rational decision for management of sustainable agriculture. Mobile phones are also used to send weather-based crop wise advisories to farmers.

Extension Methods for Development of Capacity

Climate Trainings

Innovative climate training helpful to farmers for their capacity development on different adaptive or mitigation practices, including mulching, use of organic fertilizers and organic slurry, construction of polyhouse, drip irrigation, growing of less water-intensive crops in the area, *etc.*

Climate Workshops

Climate related workshops on different topics such as preparation of water ponds, scientific cultivation of crop, application of organic fertiliser, custom hiring centre, crop insurance *etc.* conducted to make farmers aware.

Field Demonstration

Field demonstration undertaken for management of different disease in different crops. The farmers were asked to adopt proper management practices like application of organic manures, neem cake, vermicompost and biofertilisers. Similarly, the awareness has also been created regarding importance of mulching to reduce soil temperature.

Climate Smart Farmers Field Schools (CFFS)

The Farmer Field School (FFS) is a participatory, non-formal extension approach based on experiential learning that puts farmers and their demands at the centre (FAO, 2002). It is a group of 20-25 farmers from the village who meet weekly throughout growing seasons at a test field of climate-resilient technology like use of *in situ* moisture conservation. The FFS is guided by trained facilitators like extension workers, NGO workers, farmer organisation staff or trained farmers. Climate smart Field Schools is effective in advancing local-level climate smart interventions (Sala *et al.*, 2016). Climate Smart Farmers Field School in Indonesia raised awareness of climate change and promoted solutions to cope with changing rainfall patterns, such as recording and interpretation of on-farm rainfall measurements and in-field water harvesting (Winarto *et al.*, 2008).

Weather-Based Insurance

Weather based agricultural insurance enable farmers to re-invest in inputs and technologies despite during bad years. 100 per cent farmers used crop insurance scheme to protect their crops from any hazards. As per the report of CGIAR in India, 30 million farmers are covered under crop insurance schemes to protect their crops from changing scenario of climate.

Community Based Disaster Management (CBDM) Approach

The CBDM approach is helpful to build-up the capacity of communities for reduce the consequences and to cope with disasters more effectively through WOTR has developed a community-based disaster management (CBDM) approach. Overall participation of community in disaster risk reduction (DRR) is the key to minimising losses.

Agro-Meteorological Advisory Service

Agro-meteorological Advisory Service is best solution to provides location-specific agro-advisories on weather forecasting, associated agricultural advices and a phone-in help desk for farmers. For this purpose, local automated weather stations were set up in the villages and daily weather information is collected through GPS. After analysis advisory services were provided to the farmers. On the basis of this, farmers plan their agricultural activities which help them to adopt to the changing climatic conditions.

Conclusion

Extension services play a vital role in CSA because they play vital role in knowledge transfer to farmers' fields. The proper mixture of different extension approaches used in CSA largely depend on certain factors such as the complexity of extension messages, the targeted population and its geographical spread, the available technology, the type and variety of data to be collected from farmers, and lastly, on the financial means available for extension. On-the-ground implementation of extension also needs to go hand-in-hand with advocacy and awareness raising of decision makers on the imminent threat of climate change for agriculture in order to make extension more responsive to climate change and contribute to address the triple challenge of food security, adaptation and mitigation. Extension approaches with two-way information flow are particularly valuable to address climate change adaptation because they collect real-time agricultural information and are able to detect effects of climate change on a local scale that can be used for decision makers to react to threats to agriculture. There are a number of agricultural practices and technologies that enhance food security, resilience, and productivity in a sustainable manner. The proper development and technology transfer are key for the success of various adaptation and mitigation practices and technologies used in climate smart agriculture.

REFERENCES

Boa, E., Danielsen, S. and Haesen, S. (2015). Better Together: Identifying the Benefit of a Closer Integration Between Plant Health, Agriculture and One Health. In One Health: *The Theory and Practice of Integrated Health Approaches*, Chapter 22: 258-259.

Bentley, J.W., Boa, E., Van Mele, P., Almanza, J., Vasquez, D. and Eguino, S. (2003). Going Public: A New Extension Method. *International Journal of Agricultural Sustainability*,1(2): 108-123.

Cline, W. (2008) 'Global warming and agriculture'. *Finance and Development*, 45(1): 23-7.

FAO. (2002). From farmer field school to community IPM – Ten years of IPM training in Asia. Bangkok, Thailand. Food and Agriculture Organisation of the United Nations (FAO), Rome, Italy.

Franzel, S., Degrande, A., Kiptot, E., Kirui, J., Kugonza, J., Preissing, J. and Simpson, B. (2015). Farmer to farmer extension. Note 7. GFRAS Good Practice Notes for Extension and Advisory Services. Lindau, Switzerland, Global Forum for Rural Advisory Services.

FAO. (2010) "Climate Smart" Agriculture: Policies, Practices and Financing for Food Security, Adaptation and Mitigation', Food and Agriculture Organisation of the United Nations, Rome.

FAO. (2013a). Sourcebook on Climate Smart Agriculture, Forestry and Fisheries, Food and Agriculture Organisation of the United Nations (FAO), Rome, Italy.

FAO. (2013b). Climate Smart Agriculture: Source Book. Food and Agriculture Organisation of The United Nations, Rome, Italy.

FAO. (2015). Building resilient agricultural systems through farmer field schools – Integrated Production and Pest Management Programme (IPPM). Food and Agriculture Organisation of The United Nations, Rome, Italy.

Gbetibouo, G.A. 2009. Understanding farmers perceptions and adaptations to climate change and variability: The case of the Limpopo Basin, *IFPRI Discussion Paper 00849*. South Africa.

International Food Policy Research Institute (IFPRI). (2011). Value chains for nutrition. 2020 *Conference Paper No. 4.*

IPCC. (2001) Climate change 2001: The scientific basis. Contribution of Working Group I to the Third Assessment Report of the Intergovernmental Panel on Climate Change (Houghton, J. T., Ding, Y., Griggs, D. J., Noguer, M., Van der Linden, P. J., Dai, X., Maskell, K. and Johnson, C. A. (Eds.)) Cambridge University Press, Cambridge and New York.

IPCC. (2007). Climate Impacts, Adaptation and Vulnerability. Working Group II to the 41 Intergovernmental Panel on Climate Change Fourth Assessment Report, DRAFT technical summary 2006, Intergovernmental Panel on Climate Change, Geneva.

Kates, R. (2000). Cautionary tales: Adaptation and the global poor. *Climatic Change*. 45(1): 5–17.

Lipper, L., Thornton, P., Campbell, B.M. and Torquebiau, E.F. (2014). Climate –smart agriculture for food security. *Nature Climate Change* 4: 1068-1072.

Mahato, A. (2014). Climate Change and its Impact on Agriculture International. *Journal of Scientific and Research*, 4(4): 3.

Mendelsohn, R., A. Basist, F. Kogan, and P. Kurukulasuriya. (2007). Climate analysis with satellite versus weather station data. *Climate Change*, 81: 71-83.

Mloza Banda, B.C. (2014). E-adaptation to Climate Change in Malawi, *ICT Update 78*.

Nelson, G, Rosegrant, M, Palazzo, A, Gray, I, Ingersoll, C, Robertson, R, Tokgoz, S, Zhu, T, Sulser, T, Ringler, C, Msangi, S and L You. (2010). Food Security, Farming, and Climate Change to 2050. International Food Policy Research Institute, Washington DC.

Parry, M., Rosenzweig, C. and Livermore, M. 2005. Climate change, global food supply and risk of hunger. *Phil. Trans. Roy. Soc. B*, 360: 2125–2138.

Sala S, Rossi F, David S. (2016) Supporting agricultural extension towards Climate Smart Agriculture: An overview of existing tools. Global Alliance for Climate Smart Agriculture (GASCA)/FAO, Italy, P28-30.

Satapathy, S. (2011) 'Adaptation to Climate Change with a Focus on Rural Areas and India' Deutsche Gesellschaft für International Zusammenarbeit (GIZ) GmbH, India Project on Climate Change Adaptation in Rural Areas of India.

Scarbourough, V., S. Killough, D.A. Johnson andJ. Farrington (eds.). 1997. Farmer-led

Extension: Concepts and Practices. London: *Intermediate Technology Publications*

Settle, W., Soumaré, M., Sarr, M., Garba, H. and Poisot, A.S. (2014). Reducing pesticide risks to farming communities: cotton farmer field schools in Mali. Philosophical Transactions of the Rural Society.

Winarto, Y.T., Stitger, K., Anantasari, E. And Hidayah, S.N. (2008). Climate Field Schools in Indonesia: Improving "response farming" to climate change. *Leisa Magazine*, 24.

WOTR. (2009). Climate Change Adaptation (CCA) Project. Information Bulletin. Watershed Organization Trust (WOTR), Pune, Maharashtra, India.

Chapter 4

Effective Crop Residue Management Measures for Better Environment

M. Ganga Devi[1], *Ch. Anil Kumar*[2] *and K. Raja Kishore*[3]

Scientists Agronomy and Plant Protection, Senior Scientist and Head, KVK, Lam, Guntur, Andhra Pradesh

India being an agriculture-dominant country produces more than 500 million tons of crop residues annually. Basically these residues are produced in nature and are very important for survival of animals and human beings who are consumers. These crop residues are widely available on earth and act as a good source of energy or it can be converted into useful products. The wastes generated from crop have a good potential to produce energy in related energy sector. The waste produced from animal or crop called biomasses, have an interdependent relation with ecosystem. Large portions of unused crop residue are burnt in the fields primarily to clear the left-over straw and stubbles after the harvest. Non-availability of labour, high cost of residue removal from the field and risen usage of combines in harvesting the crops are the main reasons behind burning of crop residues in the fields. Burning of crop residues causes environmental pollution, is hazardous to human health, produces greenhouse gases causing global warming and results in loss of plant nutrients like N, P, K and S. Therefore, appropriate management of crop residues assumes a great significance. Long-term studies of the residue recycling have indicated improvements in physical, chemical and biological health of soil. Other possible option of crop residues management lies in utilizing a portion of surplus residue as incorporate in to soil which improve soil health, increase nutrient use efficiency and minimize air pollution. Others *i.e.* mushroom cultivation as converting of inedible crop residues into valuable food, surface mulch as conservation of soil moisture, temperature and control of weed emergence, bio-fuel and compost production. Residue decomposition in soil substantially increases the soil organic carbon and other nutrient. Systematic utilization of agricultural waste also helps to improve environmental conditions by reducing pollution caused by disposal of astronomic agricultural waste.

Keywords: *Crop residue, Recycling measures, Soil health and environment.*

Introduction

Crop residues are the biomass that remains in the field after harvesting the economic components *i.e.*, grain. These residues contains significant quantities of plant nutrients and their judicious application will have positive effect on nutrient management. Recycling the crop residues has a benefit of converting the farm waste into useful product for meeting nutrient requirement of succeeding crops. In general crops generate large amount of left over residues. Enormous quantity of crop residues are generated every year, in the form of cereal straws, woody stalks, and sugarcane leaves/tops during harvest periods. With the increase in food production, crop residues also increase which further used as animal feed, thatching for rural homes, residential cooking fuel and industrial fuel. Crop residues are known to be as a good source of organic carbon for soil microorganisms and also contribute to plant nutrients. Residue retention reduces soil erosion and runoff losses and increases the permeability of soil thus helps in conserving soil moisture. Also it helps to decrease the land preparation cost and soil evaporation. With the introduction of combine harvesters, more than 75 per cent of the rice area is harvested mechanically. However, management of the rice straw is a major challenge as it is considered to be a poor feed for the animals owing to high silica content. Combine harvester leaves behind a swath of loose rice residues, which interfere with operations of the seed drill used for planting wheat. Farmers often turn to stubble burning in order to quick the action of removing left over rice crop residue which results in huge loss of biomass and emission of Greenhouse Gas (GHG) consists of methane (CH_4) and nitrous oxide (N_2O) leads to atmospheric pollution. Stubble burning also impacts human and animal health both medically and by traumatic road accidents due to restricted visibility. In order to overcome such problems and to avoid atmospheric pollution, here in this article we will be discussing about crop residue management, residue retention along with soil health and recycling measures of crop residue.

Management of Crop Residue

Majority of land in India is used for farming and a wide range of crops (*i.e.* rice, wheat, maize, sugarcane, millets, pulses, fiber crops) are cultivated in its different agro-ecological regions. Rice and wheat cropping system produces huge quantities of crop residues and mostly uses combines for harvesting the crop. Crop residues of cereals are primarily used to feed cattle. Also rice straw and husk are used as bio-fuel and bio-oil production. Management of rice straw, rather than wheat straw is a serious problem due to presence of higher silica content when compared. This happens due to lack of proper technology transfer to recycle the residues after harvesting the crop. Farmers have several management options to convert crop residue as source of income. Profitable management practices of crop residues are the following: Animal feed, mushroom cultivation, power generation and bio-fuel and bio oil production. Also used for surface retention, soil mulching, compost making and to incorporate with soil to increase organic matter.

Crop Residue for Animal Feed

In India, crop residues of rice and wheat are primarily used as animal feed by supplementing with some additives. Being unpalatable and low in digestibility, these residues cannot form a sole ration for livestock. Also the rice straw is considered poor feed for animals due to its high silica content. Methods like physical, chemical and biological treatments have been used to weaken and break down ligno-cellulose bonds in crop residues, thereby increasing their nutritional value through which the nutritional value of rice straw can be improved. About 75 per cent of total wheat straw is used as fodder for animals and also includes operation investment for cutting into small pieces. To meet the nutritional requirements of animals, the residues need processing and enriching with urea and molasses, and supplementing with green fodders (leguminous/non-leguminous) and legume (sun hemp, horse gram, cowpea and gram) straws.

Utilization of Crop Residues for Mushroom Cultivation

Mushrooms are cultivated from a wide range of varieties of lingo cellulosic residues and also production of mushroom represents a valuable conversion of inedible crop residues into valuable food which contains 2-3 times as much protein as vegetables. Crop residues like paddy straw, wheat straw, maize straw, cotton waste, and sugar cane waste are the mostly used as substrates. However, cereals straw has gained more importance in mushroom cultivation due to its high nutrient composition. Also these residues can either be used alone or a combination of several residues to enrich nutrients within. Rice and wheat straws are known to be excellent substrates for the cultivation of *Agaricus bisporus* (white button mushroom) and *Volvariella volvacea* (straw mushroom). Straw for *Agaricus* cultivation is usually mixed with horse manure and hay and a very high conversion efficiency of the substrate into fungal bodies is possible. Rattan (2013) also showed that paddy straw alone does not provide good physical structure to compost but gave a good result when mixed with wheat straw in equal quantities.

Thermal Power Generation

Most of the institutions and departments put their effort in encouraging the usage of crop residue (rice straw) in thermal plants for the generation of electricity. Biomass, such as agricultural residue, bagasse, cotton stalks, rice husk, *etc.*, is emerging as a viable source of power for rural electrification in India. Apart from the generation of electricity for supply to state grid to meet the ever-increasing demand for energy in the state, the plant also reduces the Greenhouse Gases (GHGs) emissions. As per Cleaner Development Mechanism (CDM) estimates, the plant would supply energy equivalent of approximately 417.9 million kWh to the grid in a period of 10 years (2002–2012), thereby resulting in total CO_2 emission reduction of 0.3 million tons. The thermal plant at Jalkheri, District Fatehgarh Sahib is the first plant in India which is based on use of Biomass *i.e.*, renewable energy source. This plant can utilize rice husk, waste wood chips, straw of various plants *e.g.*, paddy, wheat, *etc.* This plant was commissioned in June, 1992 on turn-key basis by M/s BHEL for PSEB to utilize rice straw at a cost of Rs. 47.2 crores.

Incorporation of Crop Residues into Soil

Incorporation of crop residues into soil or retention on the surface has several positive influences on physical, chemical and biological properties of soil. These practices increase hydraulic conductivity and reduce bulk density of soil by modifying soil structure and aggregate stability. Also mulching with these plant residues raises the minimum soil temperature in winter due to reduction in upward heat flux from soil and decreases soil temperature during summer due to shading effect. Ploughing is the most efficient residue incorporation method through which soil can be enriched with organic matter and N, P and K contents.

Bio Fuel Production

Bio-oil can be produced from crop residues by the process of fast pyrolysis, which requires temperature of biomass to be raised to 400-500°C within a few seconds, resulting in a remarkable change in the thermal disintegration process. Condensation of biomass takes place during the process of bio oil production. Further, bio-oil is free from SO_2 emissions and produces low NO_2. Certain Canadian companies (like Dyna Motive Canada Inc.) have patented technologies to produce bio-oil from biomass including agricultural waste.

Usage of Rice Residue as Animal Bedding

Farmers have been advised to use paddy straw as bedding material for cross breed cows during winter season. Paddy straw bedding help the animals keep themselves warm and maintain reasonable rates of heat loss from the body. It also provides clean, hygienic, dry, comfortable and non-slippery environment, which prevents the chances of injury and lameness. It has been found that usage of paddy straw for bedding during winter helped in improving the quality and quantity of milk as it contributed to animals' comfort, udder health and leg health. The paddy straw used for bedding could be subsequently used in biogas plants.

Compost Making

The crop residues have been traditionally used for preparing compost. For this, crop residues are used as animal bedding and are then heaped in dung pits. In the animal shed each kilogram of straw absorbs about 2-3 kg of urine, which enriches it with N. The residues of rice crop from one hectare land, on composting, give about 3 tons of manure as rich in nutrients as farmyard manure. The decomposition process, which is hastened by a consortium of microorganisms, takes 75-90 days.

Conclusion

Crop residues are of great economic values as animal feed, fuel and industrial raw material, and in conservation agriculture for which it is a pre-requisite. The recycling of crop residues has the great potential to return a considerable amount of plant nutrients to the soil. Farmers have several management options to convert crop residue as source of income. It a great challenge for the farming community to manage crop residues effectively and efficiently for enriching the soil health and soil quality. Crop Residue Management practices should be selected to optimize crop yields with minimal adverse effects on the environment.

REFERENCES

Adam John. Alternatives to open-field burning on paddy farms, Agricultural and Food Policy Studies Institute, Malaysia. 2013; 18: 1-5.

Behera B. Recycling of crop residues for improved soil nutrient status and farm income, department of agronomy CA, OUAT, Bhubaneswar, 2018.

Ghorade I.B., Jadhavar V.R., Potadar V.R and Patil S.S Department of Environmental Science Dr. Babasabhe Ambedkar Marahtwada University, Aurangabad(India). Indian Streams Research Journal Vol - 1, ISSUE – IX October 2011 ISSN: -2230-7850

Harikrishna P. Utilization of maize stalks for mushroom cultivation and compost making, department of agricultural microbiology and Bioenergy College of agriculture Rajendranagar, Hyderabad-500 030, Acharya N. G. Ranga Agricultural University, 2013.

Kamla M, Tokkas J, Chander Anand R, Nisha Kumari. Pretreated rice straw as an improved fodder for ruminants-An overview, Journal of Applied and Natural Science. 2015; 7(1): 514-20.

Navdeep Singh, Harshdeep Singh, Sewa Singh (International Journal Of Scientific and Technology Research Volume, ISSN 2277-8616.)

Niveta J, Bhatia A, Pathak H. Emission of air pollution from crop residue burning in India. Centre for Environment science and climate resilient agriculture, Indian agricultural Research institute, New Delhi., Aerosol and Air Quality Research. 2014; 14: 422-30.

Rattan D. Use of supplements for yield improvement of *Agaricus bisporus* Lange (Sing.) Thesis submitted to the Department of Microbiology College of Basic Sciences and Humanities, Punjab agricultural university, Ludhiana, 2013.

Salar RK, Aneja KR. Significance of thermophilic fungi in mushroom compost preparation: effect on growth and yield of *Agaricus bisporus* (Lange) Sing. Journal of Agricultural Technology. 2007; 3(2): 241-53.

Singh Y, Singh B, Ladha JK, Khind CS, Khera TS, Bueno CS. Management effect on residue decomposition, crop production and soil fertileity in rice wheat rotation in India. Soil Sci. Soc. Am. J. 2004; 68: 320-26.

Chapter 5

Effect of Industrial Effluents on Soil and Water Ecosystem

E. Parameswari[1], T. Ilakiya[2], V. Davamani[1], P. Kalaiselvi[1], S. Paul Sebastian[1] and K. Suganya[1]*

[1]Department of Environmental Sciences,
[2]Horticultural College and Research Institute,
Tamil Nadu Agricultural University, Coimbatore, Tamil Nadu

Introduction

Pollution is a serious problem ever since sewage and industrial effluents are disposed into water courses and on land. It has increased with the growth of industry as well as of population. Industries can conveniently be classified in to two groups *viz.* (i) Dry process industries and (ii) Wet process industries. Dry process industries (mostly engineering and assembling industries) do not use water in their process, while wet process industries use water either as raw material or in their process 05 for both. Only a little quantity of water is absorbed in the process and the rest being discharged as effluent. Practically all the volume of water used is calculated as the volume of effluent. In addition to the process waters, large volumes of cooling water are also discharged as waste. Boiler blow down is also an effluent contributing significant quantities to the bulk. These effluents contain a wider variety of materials of both organic and inorganic nature including toxic substances and are usually discharged with or without treatment into surface waters such as rivers, streams, lakes or into oceans or on land or in to sewers. The solid wastes produced in these industries are not given much importance as they are either recycled dumped or disposed of to remote places. Liquid industrial wastes are of great concern because of their harmful effects which are described in detail in the following pages.

Effects of Industrial Effluents

(a) Effects on Surface Waters

Industries need a wide variety of raw materials and chemicals which are later discharged via their effluents. Acids, alkalis, toxic metals, pesticides and other poisonous substances such as cyanide, dyes, oils, detergents, resins, rubbers are a few to mention. Heated effluents that impart thermal loading on receiving waters and effluents containing radioactive materials are also of prime concern. Some of the effluents such as from tanning and meat packing may also contain pathogenic bacteria. The nature and extent of pollution depends on the materials present in the-effluent and on the quantity discharged. The constituents present in effluents, their effects on receiving streams *etc.*, are described in detail.

(i) Acids

Both organic and inorganic acids are discharged into rivers in the form of industrial effluents. The principal processes which contribute mineral acids are, manufacture of fertilizers, battery manufacture, brewing, textiles, mining, iron and copper pickling. Organic acids are discharged from the manufacture of rayon, fermentation plants, distilleries, dyeing industries, manufacture of leather, chemical industries *etc.* The main constituent of acid wastes is sulfuric acid. In addition, nitric acid, hydrochloric acid and phosphoric acid are also present. These acids are considered to cause damage to metals or concrete structures, pumps *etc.*, by their corrosive activity, moreover they produce hydrogen sulfide gas when they come into contacts, with the sludge and mud of the river. The lower the pH value the greater is the amount of hydrogen sulfide liberated leading to a higher degree of atmospheric pollution. Acid pollutants also destroy microorganisms and thus inhibit self-purification of the stream. Acids are reported to be lethal to fish and other forms of aquatic life.

(ii) Alkalis

Alkalis also cause damage to rivers and streams by altering pH values resulting in a break in the natural buffer system. Chemical manufacture wastes, kier liquors, wool scouring wastes, tannery wastes, cotton mercerising wastes are the principal contributors of alkalis. The alkali wastes raise the pH values even up to 12. Alkalis behave similarly as acids in destroying bacterial and microorganisms and thus inhibit the self-purification of a stream. Fish and other aquatic life are also affected by the sudden rise in pH. Strong alkalis are reported to produce asphyxiation by the coagulation of gill secretions in fish. Alums and aluminium salts have a similar action due to the hydrolysis followed by formation of hydrogen ions.

(iii) Suspended Matter

Insoluble matter in suspension is one of the commonest forms of pollution. Suspended matter is present in sewage and in most-of the industrial wastewaters. Wastes from stone quarrying works, sand washing, china clay works, contain primarily inorganic suspended solids. Sewage effluents and many coal washery effluents contain suspended solids which are organic in nature. Suspended matter, though not causing any harm is objectionable as it interferes with the self purification of streams by reducing the photosynthetic activity of Water plants and by smothering benthic organisms. Moreover, abrasive materials present in the suspended matter can choke the gills of fish and there by cause damage to fisheries. Deposition if suspended matter on stream bed destroys plant and animal life and thus the food supplies for fish are diminished. The coal washery wastewaters which are black and opaque, discolor streams and render it unfit for most uses. Suspended matters containing organic materials, will undergo putrefaction during warm weather. As a result, solids buoyed by gas may rise to the surface leading to the formation of floating masses of evil odor.

(iv) Toxic Metals

- Battery manufacturing, paint manufacturing, viscose, rayon manufacturing, electroplating, copper pickling, galvansing and rubber processing are some of the important industries which discharge effluents containing considerable amounts of heavy metals. The primary metals considered to be toxic are lead, arsenic, copper, cadmium, mercury and nickel.
- These metallic solutions are toxic and inhibit self-purification of rivers. In addition, aquatic plants and animals are also killed. If the industrial waste is discharged into sewer, metals are precipitated and settle down. Contamination of rivers with toxic metals prevents further use of water for industrial and drinking purposes. Arsenic and lead particularly make water dangerous to drink. These toxic materials are reported to cause chromosome damage and thus interfere with the process of heredity. It is reported that heavy metal ions precipitate the mucous secretion of the gills. These precipitates occupy the inter lamellar spaces and the movement of the gill filaments is arrested and respiration is prevented. Copper in about 2 parts per hundred million parts of water is fatal to sticklebacks. The lethal concentration of nickel, lead and zinc is about 1 PPM.
- In very recent *years,* a lot of attention has been given to the problems of mercury poisoning. Chloralkali industries contribute significant quantities of mercury as it is used as an electrode in the electrolysis of brine. Pulp and paper industries, plastic industries, oil refining and battery manufacture also contribute to mercury pollution. The organic compounds of mercury are toxic while liquid metallic mercury is not toxic. Mercury discharged from industrial processes is mostly in inorganic forms.
- Anaerobic bacteria in the mud at the bottom of the water course converts inorganic mercury to methyl mercury which is most injurious and concentrate in the food chain, ultimately affecting human beings.
- Cadmium ranks next to mercury m its toxicity. The principal contributor of cadmium is electroplating industry.

(v) Pesticides

Pesticides if present even in small concentrations in water are fatal to fish, crustaceans and shellfish. They are able to change the metabolic activities of the body. They also cause depression of photosynthesis of planktons.

(vi) Dyes

Now-a-days thousands of dyes are prepared and they find their way into the streams through the discharge of effluents from dyeing and printing industries, textile industries and to a lesser extent from paper and ink manufacturing industries. These dyes are prepared from coal-tar hydro carbons such as benzene, naphthalene, anthracene, toluene, xylenes *etc.*, Apart from the effluents containing chromium and iron, these dye effluents are the only effluents that impart color to water courses. These organic dyes impart color even when present in micro-quantities. Though the dyes are not harmful, they are undesirable for aesthetic reasons. Moreover they also add to the B.O.D. (Biochemical Oxygen Demand) of receiving waters. In addition to the dye material, salts of chromium and aluminium, and iron which are used as mordants are also found in dye effluents. These mordants also impart color.

(vii) Oils

Replacement of coal by oil has increased the problem of pollution. Discharge of water from engineering and metallurgical industries contribute to significant levels of pollutions by oils. Tankers, barges and boats also introduce oils to rivers, canals and lakes. The presence of oil in a water course in undesirable. It has a tendency to spread over the surface of water and the diffusion of oxygen into water is inhibited and reaeration of water is affected. The absorption of oxygen from air into water depends on the thickness of the film formed by oil. When the thickness of the film increases even above 0.001 mm, a marked effect on the absorption is observed. Oil is reported to coat the gills of fish and thus affect respiration of fish. Oil pollution of sea leads to other problems. Oil is pushed to the shore by water currents and winds. Thus beaches are spoiled and bathing becomes impossible in such places. A report from Britain indicates that sea-birds around Britain are badly affected due to oil pollution.

In addition to the above, 'water insoluble' oils, 'soluble oils' also cause serious pollution. All these soluble oils form milky emulsions with water and the formed emulsions are very stable and not destroyed by easy means.

(viii) Dissolved Solids

Chlorides, sulfates, nitrates, bicarbonates, and phosphates of sodium, potassium, calcium, magnesium iron and manganese and some of the fluorides are soluble in water. Water softening plants and drainage from salt works contribute many of these salts to river and streams. Fluorides and nitrates are considered to be toxic to man above a certain level. When waters containing excessive salts are discharged into rivers, the river course becomes brackish which affect fish life and certain vegetation. Many of the freshwater fish varieties die in salt water and this is entirely due to osmotic changes. Not only these soluble salts affect biological organisms but they also cause severe effects to pipelines, pumps and other metallic or concrete structures. Sulfates are converted either into sulfuric acid which aids which aids corrosion or into sulfides which produce odor nuisances. The conversion to sulfides is brought about by sulfate reducing bacteria.

Certain soluble salts of iron and aluminum react with the natural bicarbonate alkalinity to form insoluble hydroxides. When the river or stream is depleted of dissolved oxygen due to the formation of ferrous sulfide. Iron can react with phenols to form various colored compounds and make the streams colored.

(ix) *Synthetic Detergents*

Synthetic detergents have been developed only recently and the use of these compounds increased enormously. Synthetic detergents are used as household cleaning materials and industrial cleaning agents. A point to be noted is that natural detergents exhibit only a little toxicity. It is considered that the toxicity of synthetic detergents may also be due to its non-biodegradability. The intoxicity is completely attributed to the alkyl benzene sulfonate which is the active ingredient.

(x) *Gaseous Pollutants*

This category includes ammonia. Free chlorine, Hydrogen sulfide, ozone and phosphine. Ammonia is liberated during the manufacture of chemicals, fertilizers, and coke. Sewage and other nitrogenous organic matters contribute and increase the ammonia content to dangerous levels. Free chlorine results from the processes of textile bleaching, paper mills and laundries. Hydrogen sulfide occurs by the bacterial action and pH alterations during the discharge of viscose rayon wastes and soilfate and sulfide containing materials. Toxicity of these pollutants is well known and need not be emphasised again.

(xi) *Heated Effluents*

Nuclear power plants and thermal power stations are the prime industries the effluents of which contribute to thermal loading which leads to stratification of water course and reduction in dissolved oxygen concentration. Temperature changes affect not only fish life but the entire ecosystem.

(b) Effects on Land

- ✰ The excess content of sodium above 60 per cent and boron (above 2mg/l) are deleterious to crops.
- ✰ The high sodium alkalinity combined with salinity impair the growth of plants.
- ✰ Texture of the soil is affected by sodium and penetration of roots is prevented.
- ✰ Soil permeability is also affected by sodium, and ultimately the soil will lose its productivity.
- ✰ Suspended and colloidal impurities clog the pores and form a mat on the surface of soil preventing the passage of air, water *etc.*

(c) Effects on Sewers and Sewage Treatment Plants

- The coarse materials such as rags, feathers cover the whole surface of the bar screens of comminutors and interfere with operation. Further, they clog nozzles, over load digesters and resist pump operation.
- Acid/alkalis in the effluent corrode the treatment structures. They inhibit the biological activity and upset purification.
- Presence of sulfide leads to 'Crown Corrosion' of sewers.
- The dissolved solids form incrustation on the surface of sewers.
- Cyanide, phenols, toxic organic materials, and heavy metals such as copper, chromium (6+), and zinc suppress the biological activity as they tie up enzymes which carry out oxidation.
- Fatty materials clog nozzles, pumps, and over load digesters.
- Synthetic detergents cause foam which affects natural aeration process. The same effect is observed with oil.
- Solvents if present may cause fire hazards.
- Workers employed in sewage treatment plants may be exposed to toxic gases.

Chapter 6

Resource Conservation Technologies (RCT's): A Viable Option to Combat the Climate Changing Scenario

Jitendra Singh Bamboriya[1], Mounica Talasila[2], Somdutt[3], Neeta Mahawar[4], Pratishtha Diwaker[5], Swetha Dhegavath[6], Tirunagari Rupesh[7] and Ganga Ram Mali[8]*

[1]*Ph.D. Research Scholar (Soil Science and Agricultural Chemistry), MPUAT, Udaipur, Rajasthan*

[2]*Graduate Student, Hutson School of Agriculture, Murray State University*

[3]*SRF, Directorate of Research, SK Rajasthan Agricultural University, Bikaner, Rajasthan*

[4]*Ph.D. Research scholar (Soil Science and Agricultural Chemistry), RVSKVV, Gwalior, Madhya Pradesh*

[5]*Ph.D. Research Scholar, Department of Horticulture, RCA, MPUAT, Udaipur, Rajasthan*

[6]*M.Sc. Research Scholar, Department of Soil Science and Agricultural Chemistry, PJTSAU, Hyderabad, Telangana*

[7]*Ph.D. Research Scholar, Division of Soil Science and Agricultural Chemistry, ICAR-IARI, New Delhi*

[8]*Program Assistant, KVK, Gudamalani, AU, Jodhpur, Rajasthan*

Millions of global poor people, who are already suffering from food insecurity, are expected to be impacted further from biotic and abiotic stresses arising from current and future projected climate change. By 2050, global food demand is likely to increase by ~60 per cent above 2006 levels to feed about 9.1 billion populations requiring additional production of one billion tonnes of cereals. Climate is an important biophysical factor influencing the growth, development and productivity of crops. The climate change would lead to more frequent extreme events, floods, droughts, and cyclones which in turn, would result in greater instability in food production impacting poor and marginal sections of the society. Climate change coupled with intensive unsustainable crop production practices is exerting pressure on natural resources resulting in stagnant or declining productivity per unit of land area or/and applied water, nutrient, labour, and other input resources even after the adoption of good management practices. There is urgent need to make concerted efforts for the eradication of global hunger, poverty and malnutrition while addressing the challenges of declining natural resources amidst growing threats of climate change and variability. Conservation and efficient utilisation of natural resources holds the key for adapting to the variable climate which has co-benefit of reducing the GHG emissions. Adoption of resource conservation technologies (RCTs) based on conservation agriculture (CA) and conventional agriculture in place of non-RCTs is essential for the efficient use of natural resources. This Chapter focuses on the RCTs which can contribute to efficient use of resources, adaptation to climate change and also minimise GHG emissions

Keywords: *Adaptation, Climate change, Conservation agriculture, Conventional agriculture, Resource conservation technologies.*

Introduction

Climate is an important biophysical factor influencing the growth, development and productivity of crops. Seasonal climate variability and increase in frequency and intensity of extreme weather events are important environmental threats of the 21st century (Edame *et, al* 2011). Climate change is often used interchangeably with global warming. Climate change is a significant change in temperature, precipitation, or wind lasting for an extended period (decades or longer) and it caused by both natural and anthropogenic factors. The anthropogenic intervention in the global climatic system through release the concentration of greenhouse gases (GHGs) into the atmosphere has led to serious concerns because of their induced impacts which include climate change and climate variability.

The increase in global atmospheric temperature and carbon dioxide (CO_2) levels over the long-term average are important indicators of climate change. The global mean temperature has increased by 0.74°C during the past 100 years and by the end of 21st century, it is expected to rise by 1.84° C As per the IPCC (2019) report, in South Asia, the rise in mean temperature is projected to be 0.88- 3.16° C and 1.56-5.44° C by the end of 2050 and 2080, respectively.

Likewise, the global atmospheric CO_2 concentration continued to increase since pre-industrialization time (1750) from approximately 277 ppm in 1750 to 315 ppm in 1967, and reached 412 ppm on an average at the end of 2019, which is nearly 47 per cent above the pre-industrial period. The greenhouse gas (GHG) concentrations show that CO_2 levels continued to increase during 2015-2019, with growth rates of nearly 20 per cent higher than the previous five years. India's annual GHG emissions from agriculture and livestock stood at 417 Gg CO_2 during the year 2014, of which 42 per cent was emitted from crop production and burning of crop residues, and 54 per cent came from livestock.

Agriculture is one of the sensitive sectors to the growing climate change and variability. While there are uncertainties on the nature and extent of the impending changes, climate-induced variability and change will certainly affect the agricultural production and food security. Climate change effects directly and indirectly on land degradation, food production, crop productivity, water supply, and environmental hazards. The direct effects are through changes in agroecological conditions and indirectly through affecting individual's income and demand of food produce (Kumar *et al.*, 2018).

Emission from the ruminants is the largest source, followed by NO_2 emission from soils (19 per cent), by rice cultivation (18 per cent), and about 7 per cent from manure management (MoEFCC, 2018).Numerous studies indicated that the direct impact of climate change and variability on Kharif crops is significant as crops during the Kharif season are more vulnerable to increased extreme weather hazards such as decreased rainy days and rainfall intensity, increased duration and intensity of drought and floods, changes in atmospheric humidity, diurnal asymmetry of temperature, pest incidence and virulence. Similarly, Rabi crops are comparatively more susceptible to an increase in average minimum temperature, and asymmetry of day and night temperature.

Acclimatization and Elevation Management Practices to Combat Climate Change Scenario in Agriculture

Acclimatization and elevation are often viewed as two separate activities, the former aiming to adjust to the expected increase in temperature and to the associated impacts and the latter aiming at reducing the GHG emissions. The IPCC denies adaptation as 'the process of adjustment of agricultural systems to the actual or expected climate and its effects'. Adaptation is achieved by enhancing adaptive capacity, or by minimising exposure and sensitivity to the harmful effects of climate change. Elevation of climate change consists of strategies or technologies in agriculture and allied fields which can contribute to minimize the GHGs emissions and increase the carbon sequestration of soils through changes in land-use management and enhancing input-use efficiency. India has the potential to cut its annual GHG emissions from agriculture and livestock, which is equal to 85.5 mega tons of CO_2 equivalent per year. It could be achieved with minimal cost by efficient use of fertilizer, adoption of zero-tillage and water management. Some important crop and soil management strategies for reducing greenhouse gas emissions and increase carbon sequestration are the use of different methods of rice cultivation like mechanical transplantation, SRI, DSR, improved cultivations such as direct seeding of the corn, wheat and maize FIRB system; no-tillage system; precise land levelling; precision water management approaches including drip and sprinkler systems; and precision agriculture including sensor based reliable nutrient management which can rationalise fertilizer application and minimize excessive use (Srinivas Rao *et., al.* 2016).

By shifting water management from the current practice to intermittent flooding in all the irrigated rice growing areas of the country, the CH_4 from irrigated rice fields could be reduced by 40 per cent. It was also stated that by modifying water management from continuous flooding to alternate flooding or applying urea alone rather than urea coupled with FYM would reduce global warming potential by 15 and 29 per cent, respectively. Similarly, SRI and DSR decrease CH_4 emissions drastically having a considerable potential to reduce the global warming potential by 35-75 per cent over the conventional transplanted rice. The need-based nitrogen application using leaf colour chart (LCC) has the potential to reduce $N0_2$ emission and global warming potential by around 11-14 per cent. Such options that capture the atmospheric carbon and store it in the soil are important not only for global warming mitigation but also can contribute towards improving the soil fertility status (Srinivas Rao *et al.*, 2017). Moreover, the GHGs emissions can be abridged by the substitution of fossil fuels by the agricultural feedstock (*e.g.*, crop residues, dung, and dedicated energy crops) for energy production.

Resource Conservation Technologies (RCT'S): A Productive Approaches for Conservation and Intensive Agriculture

In recent times, the agricultural production is almost stagnating and factor productivity is declining even under good management and with full recommended doses of all inputs due to degradation of natural resources including land, water, nutrient, energy, labour, carbon, *etc.* This is more evident for rainfed agroecosystem (Srinivas Rao *et al.*, 2016). This is often associated with biotic and abiotic stresses, excessive use of agrochemicals, unavailability of qualitative germplasm, and lack of holistic approaches to address the situation and problem.

The intensive cultivation without considering for sustainability of agro-ecosystem resulted in the reduction in carbon level since long. The intensive cultivation leads to reduction in the soil carbon stock, though the extent of decrement differs with the ecosystem and management practices such as soil cover, climatic and edaphic characteristics, and farming practices. The global agricultural soils have already lost organic carbon by 3075 per cent from their antecedent soil carbon because of intensive cultivation (Meena *et.al.*, 2020). The cultivated soils of India severely exhausted the soil organic carbon stock which ranged from <1.0 g Kg^{-1} to hardly 10-15 Mg C ha^{-1} in upper 4 cm soil Horizons. (Lal, 2015). Technologies like minimum tillage and crop residue management, and cover crops primarily focus on resource conservation, savings, ensuring soil nutrients and moisture conservation and adoption of spatial and temporal crop sequencing. RCTs bring many possible benefits including reduced water and energy use (fossil fuels and electricity), reduced GHGs emissions, soil erosion and degradation of the natural resource base, increased yields and farm incomes, and reduced labour shortages (Rani *et.al.*, 2019).

Guarantee Based Assuring Resource Conservation Technologies (RCT'S) Recommending for Climate Change in Agriculture

Conventional Agriculture Based RCT'S

A. Precision Land Levelling

Precision land levelling includes smoothening the land surface by using laser equipped drag buckets and it creates a slope of 0 to 0.2 per cent. It involves the use of large horsepower tractors equipped with laser-guided instrumentation or global positioning system (GPS) for creating the desired slope and it allows uniform water distribution with negligible losses by providing a very graded and smooth field condition. It facilitates uniformity in seed and fertilizer placement which helps to maintain optimum plant stand, enhanced nutrient acquisition and higher yield. Review suggests that laser land levelling resulted in reducing irrigation water application by 25 per cent and labour requirements by 35 per cent, and an increase of net area by 3-4 per cent and wheat yield by 30 per cent as compared to conventional practices of field preparation. Thus, by increasing water use efficiency it puts less pressure on the existing water resources and helps to adapt to variable moisture environments.

B. Crop Establishment Methods in Rice

In transplanted rice, puddling requires 30 per cent of the total water requirement which is a major constraint under depleting ground water resources along with the climate change and labour shortage. This necessitates rice cultivation by various methods like DSR, bed planting for reducing environmental impact compared to conventional transplanted method.

System of Rice Intensification (SRI)

SRI is a worthy technology for resource poor farmers is being adopted in significant scale in major rice-producing countries like Vietnam, China, India, and the Philippines. Moreover, researches indicate rice under SRI to be hardier against extreme weather situations and biotic stress due to improved root strength and plant vigour. Alternate irrigation to the crop aims at minimizing the water use, whereas wide spacing and early transplanting contribute to tillering. Rice contributes about 15-20 per cent of the total methane emission that is generated by anthropogenic activities. Through alternate wetting and drying, SRI creates aerobic soil conditions which reduce methane emission. Researches indicate that intermittent application of irrigation by SRI is lessens methane emission by about 22 per cent (Chu *et al.*, 2015). However, it has been identified that higher labour requirement is a major limitation restricting its adoption.

Furrow Irrigated Raised Bed Plant System (FIRB)

This technology is widely adopted in the Indo-Gangetic Plains, it is an efficient conservation technology and a potent method to conserve water (30-50 per cent) (Kukal *et al.*, 2005), reduce seed rates, facilitate mechanical weed control, minimize wheat lodging (Chauhan *et al.*, 2015) reduce expenses and conserve resources. Band application of fertilizers can be done easily through the crop residues, reducing thereby the nutrient losses under permanent raised bed planting.

Moreover, this planting technique provides an opportunity for inclusion of different crops, thereby opening avenues for generating alternate sources of income and productivity through efficient use of resource base in the context of climate change. In wheat cultivation, FIRB results in 25-40 per cent saving in seed, 25-40 per cent saving in water and 25 per cent reduction in nutrient application without compromising yield. This method of planting is best suited for wheat as it is adapted to close planting geometry (Das and Yaduraju, 2011).

Direct Seeded Rice (DSR)

Shortage of labour, water, timeliness, and increased methane emission are some of the reasons behind the shift from transplanting to DSR (Pandey and Velasco, 2005). DSR yields similar to transplanted rice when weeds are managed below the economic threshold level with judicious herbicide use. Further, labour requirement can be reduced by 50 per cent and water saving can be achieved to the extent of 35-57 per cent as compared to transplanted rice. In the IGP, water use was reduced by 12-60 per cent under direct-seeded, transplanted rice grown on raised beds in comparison with flooded, transplanted rice. Hence, DSR can be a mitigation strategy for meeting the water demand of rice under higher temperature as for each degree temperature rise, water demand increases by 2-3 per cent. Besides, seeding rice directly into the fields rather than transplanting into flooded paddies would curtail methane emission and hinder the process of global warming. Under wet-DSR technique, methane emission reduction was increased from 16-22 per cent (continuous flooding) to 80-90 per cent (midseason drainage) as compared to conventional techniques. Studies shows lower values of Greenhouse gas index (GHGI) from DSR, which indicates that DSR is more advantageous in mitigating greenhouse gas emission and producing more yields. Therefore, in the wake of climate change, DSR is an economically and technically viable alternative to conventional transplanted rice.

Mechanical Transplanting

Rice transplanting in traditional practice is labour intensive and a drudgery operation. The reduction in man power availability and increased availability of mechanical transplanters made a paradigm shift towards the mechanisation in rice farming, and the Government is making sufficient efforts in promoting mechanisation to make the farming profitable. Transplanting mat type nursery is mostly used in mechanical transplanter due to its higher performance and less man power requirement of 50 man-hr/ha (Dixit *et al.*, 2017). In spite of having advantage over the traditional transplanting, adaptability of mechanical transplanters by farmers is low due to high initial investment and lack of technical knowledge in growing mat type nursery. Improving technical knowledge and availability of machinery in time through custom hiring are practical solutions for increasing the mechanical transplanting in rice.

C. Intercropping

Intercropping is an adaptation strategy on spatial and temporal perspective and functional diversication of different crops, and it provides better utilisation of locally bound agroecosystem resources. It is an ancient practice which represents within-eld ecological intensication and diversication strategy maximizing the land productivity. This increase in productivity contributes to accumulation of soil organic matter in higher side and induces carbon sequestration to mitigate greenhouse gas production from agriculture. Sowing of crops along with legumes provides stable yields by efficient utilization of resources, reduces variability in crop yields and improves soil quality through atmospheric fixation of nitrogen (Layek *et al.*, 2018). High leaf cover provides cooler microclimate through transpiration which in turn, reduces soil evaporation and associated soil temperatures (Miao *et.,al* 2016).

D. *In situ* Moisture Conservation and Water Harvesting

Water harvesting, water conservation and improvement in irrigation accessibility and efficiency are need of the hour for boosting productivity and livelihood improvement, under changing precipitation patterns and elevated temperatures. For better farm management, farmers have to be trained and motivated for adopting micro-irrigation systems and on-farm water conservation techniques. Thus, methods involving decreasing runoff (contours, vegetative hedges), increasing water infiltration and reducing soil evaporation (crop residues mulch) should be advocated for better soil-water management.

Water harvesting is a process of collecting runoff water into a catchment for further use. The collected water can be directly applied to the crop or stored in the soil role for immediate crop use (runoff farming), or stored in an on -farm water tank for future uses. Areas where water is scarce due to low rainfall, simple techniques for *in situ* conservation, groundwater recharge, farm pond or tank and protective irrigation are important to minimize the impact of dry spells and for providing opportunities for critical irrigation. *In situ* water harvesting techniques include summer ploughing, random tie ridging, broad bed and furrows, ridges and furrows, and compartmental bunding, whereas *ex situ* techniques include creation of farm ponds (Srinivas Rao *et al.*, 2017). The primary objective of *in situ* conservation is to control soil erosion and ensure minimal runoff.

In the case of drought or erratic rainfall, *ex situ* water management *i.e.*, collection and storage of rain water in ponds will provide convenient water supply during dry periods through supplementary irrigation (Tolossa *et al.*, 2020). Collection and storage of rainwater can provide a convenient and reliable water supply during seasonal dry periods and droughts. Rainwater collection can also contribute greatly to the stabilization of declining groundwater tables, decrease travel time to remote water sources and increase agricultural productivity.

E. Mulching

Mulching involves covering the soil surface with organic or inorganic materials, which helps to moderate soil temperature and moisture, and reduce nutrient losses, salinity, erosion, *etc.* In the wake of climate change, elevated temperature, soil erosion, landslides, *etc.*, mulching has regained its importance. Application of crop residue mulch at the soil air interface protects the soil against raindrop impacts, decreases runoff velocity, increasing infiltration rate and also acts as a sediment trap. Besides, mulch has favourable impact on soil quality and resilience; it moderates soil temperature and moisture regimes and has beneficial effect on crop growth and yield. Furthermore, it is quite probable that mulch enhances soil organic matter content, aggregate stability and microbial activities.

F. Weed Management

Climate change has a serious impact on the interaction between crop and weed and thus, it is necessary to understand the weed interaction and dominance pattern for successful weed management. Under increased temperature scenario, weeds would pose serious 4 yield penalties. In the case of C crops 3 like rice and wheat, weed problem will aggravate due to increase in CO_2.

Some weed species like *D. aegyptium* and *E. colona* responded to elevated CO_2, whereas *C. rotundas* and *E. indica* hardly responded to carbon enrichment. Besides weed shift, problems like herbicide toxicity and resistant weeds may increase as a result of continuous herbicide use and climate change. Integrated weed management (IWM) is the more applauded aspect of long-term weed management programme under climate changing scenario as it brings all feasible weed control methods into a single and co-ordinated system (Ghosh *et al.*, 2017). Effective utilization of allelopathic crop residue mulches (wheat, barley, maize, sorghum, rye) may help to mitigate adverse climatic effects by acting as self-supporting weed management method for the concurrent as well as succeeding crops. Biological weed control methods (use of bioagents, bioherbicide) usually target the specific natural enemy weeds and thus, can be a promising method to naturally limit the weed population. Soil solarization plays a vital role in weed, nematodes and pathogen management under increased temperature conditions.

G. Nutrient Management

Optimal utilization of on-farm resources as inputs such as crop residues; location and time-specific application of major nutrient fertilizers according to the needs of plant; selection of most appropriate combination of fertilizer sources for enhancing use efficiency and economic benefits; integration of nutrients from different sources along with other integrated crop management practices for sustainable crop production; and correction of deficient nutrients and rationalizing nitrogen fertilisers are some of the strategies, which can minimize the resource use and enhance efficiency, and also minimize emissions.

Biochar

Lehman *et al.*, (2003) reported that there is a reduction in soil nitrogen leaching by 60 per cent; reduction in 20 and 10 per cent of fertilizer and irrigation applications, respectively; and 38-45 per cent increase in productivity of crops through application of biochar. Biochar application to soil acts as a refuge site for microbial colonization and gets protected against grazing by their natural predators.

Brown Manuring with Sesbania

Due to scarcity of water during summer season growing of green manure crop like Sesbania and incorporating it in the soil is a big challenge now-a-day. It requires additional irrigation water and fuel costs during incorporation, and farmers are unable to tap the full potential of green manuring during the late onset of monsoon. Brown mulch/manuring provides quick coverage of surface in inter-row spaces used for moisture conservation and improves soil organic carbon content and can be considered to be one form of carbon farming. Brown manuring with daincha in drum seeded rice has significantly reduced weed density, accumulated dry matter and increased rice yield.

H. Change in Land-Use Management Practices/Agroforestry

Since agriculture is contributing to greenhouse gas production directly at the global level, integration of trees in agricultural cropping systems as agroforestry is a viable mitigation strategy for climate change. Agroforestry systems include horizontal crop diversification, crop rotation systems, border planting, hedgerow intercropping, live fences, and a mixed strata agroforestry with different land management practices. It has immense opportunities in the climate changing scenario by producing assets for farmers along with promotion of sustainable production which in turn, improves agro ecosystem diversity and climate resilience. Agroforestry practices have positive social and environmental impacts, which receive more attention in the global agenda on climate change mitigation. It requires advanced cultivation practices where trees and animal components are added to the cropping systems which ensure swift adoption (Matocha *et., al,* 2012). Improved altered microclimates in the agroforestry system provide extra resilience to shift in temperature and moisture regimes. This system increases farm net income through improvement and diversification of farm output by efficient utilization of farm resources against natural calamities and adds to financial diversity and flexibility of overall farm enterprises which contribute to climate change mitigation. Altered land use patterns such as shift in the production of crops, animal husbandry, and change in the application intensity of inputs help to improve risk bearing capacity of the farmers from climate change. Furthermore, agroforestry also has possibilities, like enhancing total (biomass) production, increasing soil fertility, improving soil health, nutrient cycling, microclimate change, carbon sequestration, bio-drainage, and biofuel production.

I. Better Weather Forecasting and Crop Insurance Schemes

Weather forecasting and forewarning systems play a pivotal role in minimizing the risks associated with climate change. In this context, information and communication technologies (ICT) can be of immense help in the dissemination of weather forecast and weather-based agro advisory. Crop insurance schemes help farmers in tackling the risk of crop failure due to climatic variability and extreme events. Weather Based Crop Insurance Scheme (WBCIS) reduces the distress of the insured farmers against the monetary loss resulting from adverse climatic conditions, whereas yield-based insurance compensates farmers against crop yield loss. The recently introduced Pradhan Mantri Fasal Bima Yojana is a comprehensive insurance scheme for minimising the risk due to climatic vagaries, loss due to pests, *etc.* Thus, there is an urgent need for the development of sustainable insurance system and farmers need to be trained for availing such opportunities.

Conservation Agriculture Based RCT'S

During the early days, farmers used to loosen the soil with a stick for easy sowing of crop plants. With the advent of iron implements, the "tillage" got more importance due to its numerous benefits. However, excessive tillage led to a dust bowl in the USA during the 1930s. This paved way for conservation agriculture (CA), which got more attention with raising concerns of resource degradation and high input costs during the post green revolution era. Conservation tillage is the collective umbrella term widely used for any practice with a conservation purpose including zero/no tillage, minimum tillage, reduced tillage, *etc.* CA seeks to combine the management of all agricultural resources including soil and water in order to achieve sustainable agricultural production.

Therefore, it can also be referred to as resource efficient or resource effective agriculture. The FAO defines the Conservation Agriculture (CA) as a farming system that promotes minimum soil disturbance, maintenance of a permanent soil cover, and crop diversification for enhancing above and below the ground biodiversity and natural biological processes to achieve improved water and nutrient use efficiency and sustained crop production. Three basic principles of CA are, i) the disturbed area should not be more than 15 cm in width or 25 per cent of the total cropped area, ii) at least 30 per cent of the land area must be permanently covered with organic materials and iii) Crop diversification should include at least three different crops in a year.

A. Minimum/No Disturbance of Soil

Usually, tilling of the soil is done to prepare a good seedbed for easy sowing of crops and to manage weeds in the crop field. However, excessive tillage leads to the destruction of soil structure and stimulates the germination of weed seeds by bringing them to the soil surface (Das, 2008). Following crop establishment practices are usually followed under CA.

Zero Tillage/No Tillage

In which the primary tillage is totally avoided and the secondary tillage is restricted to the crop row zone only. This technology mostly used in Rice-Wheat cropping system using Zero till-ferti-seed drill machine, which is capable of placing the seeds and fertilizers in narrow slits created by a furrow opener. This technology reduces the number of tractor operations required for field preparation and crop sowing and thus, saves fuel, energy, cost and time. Several reports stating a saving of 70-88 per cent in tractor hours and 67-89.5 per cent in fuel consumption. The sowing of zero tilled wheat after rice can be advanced by 7-10 days as compared to traditional methods, thus avoiding terminal heat stress in wheat. It has been reported that zero tillage resulted in a two per cent.

Minimum Tillage/Reduced Tillage

The primary tillage is restricted to the crop row zone only in reduced tillage. It has no apparent benefits over zero tillage; but farmers are sometimes forced to go for reduced tillage instead of zero tillage due to the following reasons (Sharma *et al.*, 2007). i) Considerable time gap between early harvest of rice and sowing of subsequent wheat crop. ii) Wet soil due to late irrigation/rain towards the end of rice season leads to the formation of tracks after combine harvesting, which makes the field uneven. iii) The chemical control of some broad-leaved weeds like Rumex sp. (that germinates in October after the harvest of rice crop) becomes costlier than minimum tillage.

Direct Seeding of Rice (DSR)

DSR in dry unpuddled soil can be done without any primary tillage. It is a labour, fuel, time, water, and energy saving technology that emits less greenhouse gas as compared to transplanted rice. Pathak *et al.* (2013) reported a saving of 3-4 irrigations in DSR as compared to the transplanted rice (TPR) without any yield penalty. There was also a saving in human labour use (45 per cent) and tractor use (58 per cent) in the DSR as compared to TPR. In various districts of Punjab, the GHG emissions under TPR was reported to be between 2.0 and 4.6 CO eq/ha, whereas under DSR it ranged from 1.3 to 2.9 CO_2 eq/ha.

Permanent Bed Planting

Permanent beds are prepared by tilling the soil for the first time, which can be used in subsequent years with only a little reshaping of the beds. In this technology, 2 to 3 or 4 to 5 rows of a crop can be grown on a raised bed of 60 cm (narrow bed) or 120 cm (broad bed), respectively. A furrow of 30 cm depth separates two adjacent beds. The bed planting results in saving of water (~30 per cent), seeds, nutrients, weed suppression and better solar radiation interception.

B. Permanent Soil Cover

In conservation tillage at least 30 per cent of the soil surface is covered with crop residues. Leaving crop residues on the soil surface without incorporation has numerous benefits such as soil and water conservation, weed suppression, buildup of soil and organic carbon, slow release of nutrients, *etc.*

Seeding in to Surface Residues

Various machines are available for sowing seeds in standing surface residue and also in loose residues which normally exists when the combine is used for harvesting paddy. However, the amount of crop residue that can effectively be managed by surface retention is a matter of concern. The traditional zero-till-seed drill can operate under surface residue of up to only 3 t/ha. 'Happy Seeder' can plant wheat seed under a surface residue of up to 10 t/ha. A super straw management system (SMS) can be attached to self-propelled combine harvesters during harvesting of rice which cuts paddy straw into small pieces and spread it on the soil surface, which makes sowing of wheat easier using a happy seeder.

Soil Properties under Surface Residues

About 95 per cent of total soil erosion occurs due to the detachment of soil particles by the impact of raindrops, while the rest five per cent occurs due to runoff. Therefore, the ideal solution for soil conservation is the maintenance of soil covered mulch with growing plants or surface residues instead of physical barriers, which usually controls only runoff. Residue retention increased the water-stable aggregates of >2 mm and 0.25–2.0 mm size by 23 and 10.1 per cent over residue removal, respectively (Singh *et al.*, 2016). The residue acts as a barrier against the impact of raindrops and increases the infiltration opportunity time by reducing the energy of droplets. Seeding in to Surface Residues that the surface of no-till soils resulted in about 35 per cent higher potentially mineralizable N compared to conventional till soils. It was due to greater soil microbial population and enzymatic activities and thus, greater conservation of N in CA plots. A stable soil structure is created due to the accumulation of organic matter resulting from increased biological activity. The pre-harvest litter fall from pigeon pea acts as a mulch which can add up to 90 kg N/ha. This N is mineralized slowly during the subsequent season, thus become available to the next crop. CA improve soil pH, organic matter and available nutrient contents. CA has the potential to sequester atmospheric CO_2 in the form of soil organic matter and enhance soil fertility when crop residues are used as mulch. The appropriate use of deep-rooting cover crops leads to nutrient recycling and reduced runoff that results in higher nutrient use efficiency. The better placement of fertilizer with the seed drill under CA led to improved fertilizer use efficiency (10-15 per cent) in the rice-wheat system as compared to broadcasting with the traditional system.

Weed Management under Surface Residues

The weed dynamics change significantly over the years under surface residue condition. The annual weeds get very problematic during the initial period of zero/ minimum tillage practice till the exhaustion of weed seed bank in surface soil (0-2 cm). After the exhaustion of weed seed banks from surface soil, annual weeds get suppressed significantly as deep-buried weed seeds do not get a chance to come up to surface soil. Additionally, the decomposition of surface residues secretes allelochemicals, which inhibits the germination of both annual and perennial weeds. Under continuous zero tillage conditions, perennial weeds may become very prominent due to their vegetative propagating materials. These perennial weeds can be controlled by using suitable chemical herbicides. Brown manuring is a technology that can be used to manage weeds under conservation agriculture by using leguminous plants as both live and dead mulch. It is a practice of growing leguminous plants like Sesbania along with the main crop for a period of 20-30 days, after which a selective herbicide like 2,4- D or Bispyribac-Na can be sprayed to knock-down the brown manure species. After the knock-down, the dead residues of brown manure species can suppress the weeds. Although it is a common practice in rice, it can also be extended to maize (Behera *et al.*, 2018) and pearl millet.

C. Crop Diversification

The rice-wheat cropping system of IGP remains fallow for 60-70 days between the harvest of wheat (April) and sowing of subsequent rice crop (July). The short-duration moon bean varieties like SML 832 and SML 668 can be grown during this fallow period. The inclusion of legumes in the cropping system acts as a break crop for various pests. The crop rotation with legumes meets the N requirement through biological fixation of atmospheric N.

D. Site Specific Nutrient Management

Next to seeds, fertilizers are important input for food production. The overuse of inorganic fertilizers has caused major environmental problems with an estimated 60 per cent nitrogen pollution caused by crops alone. Fertilizer-use efficiency (FUE) in South Asia is very low, *i.e.*, 30-40 per cent (Tewatia *et al.*, 2017). By enhancing the FUE by 5 to 10 per cent, it could increase returns, reduce emissions of N O and significantly reduce production costs, thus giving greater profits. Site specific nutrient management is a modern approach for managing specific resources dependent on field conditions that can save precious inputs and costs. It can be used not only on large mechanized farms but also on small farms in the tropics. In which nutrients are applied in the right form, in the right amount and in the right place to improve fertilizer use efficiency. These basic principles of precision nutrient management are applicable to fields over 1000 ha or less than 0.5 ha. Leaf N concentration is linked with leaf chlorophyll content; therefore, leaf greenness can be used as the basis of in-season nitrogen management (Bijay *et, al,* 2015).

Chlorophyll meter (SPAD), leaf colour chart, and NDVI green Seeker are currently being used in farmer's fields. Similarly, Nutrient Expert and Crop Manager, computer/android phone-based decision support systems are progressively being used to enable the application of improved nutrient management practices in farmers' field. In IGP 'nutrient expert'-based nutrient management reduced the GHG intensity of rice, wheat, and maize production by 5-35 per cent (average 13 per cent) over farmers' fertilizer practices based on CIMMYT studies. Customized leaf colour chart (CLCC) is being used for real-time N management in rice, wheat and maize. Nitrogen is applied in synchronisation with crop N demand and thus, helps to reduce the total nitrogen use and GHG emission. According to Bhatia *et al.* (2012) -1 application of 120 kg N ha at LCC4 in rice reduced emission of nitrous oxide and methane by 11 and 16 per cent, respectively which in turn, reduced -1 the GWP by 1297 kg CO ha in LCC4 2 than the conventional urea application.

E. Precision/Climate Smart Water Management

With rising temperature and changes in precipitation patterns due to climate change, it is likely to have serious implications on water availability and thereby, affecting agriculture. There are certain technologies like DSR, SRI, aerobic rice cultivation, alternate wetting and drying, deficit irrigation, partial root drying, intermittent irrigation and micro-irrigation management are well proved for enhancing water use efficiency as well as to cope with challenges of climate change and variability. Alternate wetting and drying (AWD) is a water management technique in rice, which was developed in the 1970s to address the problem of increasing water scarcity in agriculture. Richards and Sander (2014) disclosed that 20-70 per cent reduction in methane emissions along with 25-40 per cent saving of water as compared to conventional ooded rice cultivation. Therefore, AWD serves the twin purpose of mitigating GHG emissions by reducing methane and adapting to water stress. Similarly, SRI, ARS and DSR save 20 to 50 per cent irrigation water in rice cultivation. Micro irrigation is another vital option for high irrigation efficiency (> 90 per cent). However, majority of the area (55.4 per cent) covered under micro irrigation is in horticultural crops and only 7 per cent area is under field crops.

Sidhu *et al.,* (2019) evaluated the sub-surface drip fertigation (SSDF) in rice-wheat cropping system under conservation agriculture and showed saving of irrigation water by 48-53 and 42-53 per cent in rice and wheat, respectively under combination of SSDF and CA compared to flood irrigation system. The SSDF system limits evaporation from the soil and allows better delivery of water and nutrients directly to the root zone, which leads to efficient fertilizer application, discourages weed growth and reduces labour cost, and it is easier to perform cultural farming practices in SSDF compared to surface drip irrigation. Since energy requirement for pumping of water using drip irrigation (SDF or SSDF) is 50 per cent lower than that of flood irrigation, it can be met through solar powered pumps and hence, reduces the water footprints significantly.

Elevation Strategies

The GHG emissions from agriculture sector in the year 2014 amounted to 16 per cent (417.22 Gg t CO_2) of the gross emissions in India, which includes 74 per cent methane (CH_4), and 26 per cent nitrous oxide (NO). GHG emissions from various sub-sectors of the agriculture sector are illustrated in Figure 6.1. Emission of CH_4 mainly occurs in this sector due to 4 animal husbandry (enteric and manure management) and cultivation of paddy. The application of nitrogenous fertilizers on agricultural soils primarily emits NO (MoEFCC, 2018). Crop-specific management practices need to be adopted in order to further minimize emissions from the agriculture sector. Better crop management practices increase resource efficiencies that often reduce GHG emissions. The extent of adoption of such practices depends on factors like people's readiness to change, soil type, climate, farming systems, and input support.

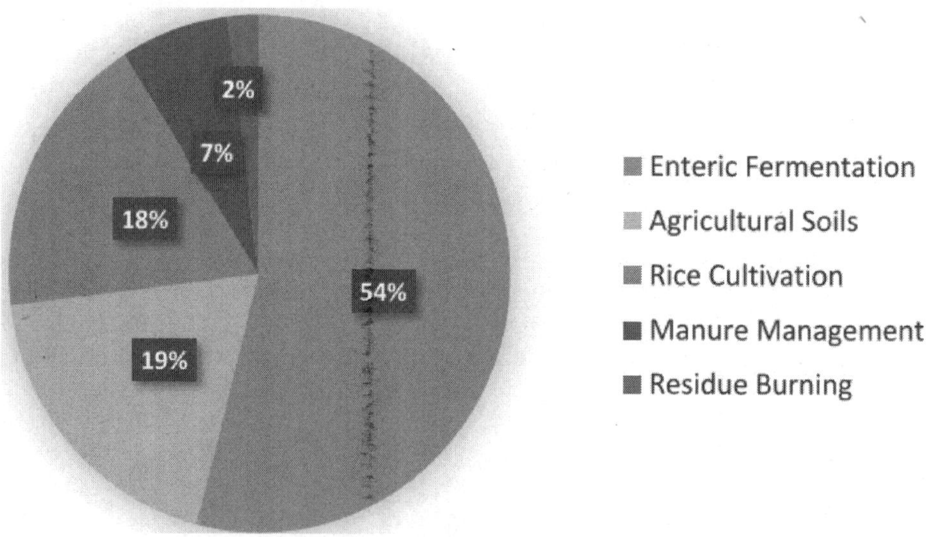

Figure 6.1: Sub-sector-wise Distribution of GHG Emissions from the Agriculture Sector.

GHG Elevation Potential of Ca-based Management Practices

No-till/reduced till leads lower exposure to oxidation of organic soils and lower atmospheric CO_2emissions, particularly in comparison with tilled soils.

Various conservation agriculture management practices are followed in irrigated and rainfed based cropping systems for the reduction of GHG.

A. Elevation Potential of Irrigated Ca-based Cropping System's

Research indicates that India's rice-wheat system can sequester 44.1Gg tonne carbon in just 20 years, if it is transformed into no-till. It is also estimated that in the same period of time if no-till based cotton-wheat and maize-wheat system adopted in India can sequester 6600 Gg C (Aryal *et al.*, 2015). The major source of methane emission is conventional transplanted rice. The introduction of DSR is a viable alternative for reducing agricultural GHG emissions. By the adoption of DSR instead of transplanted rice in the Punjab state, the global warming potential of paddy crop could be reduced by 33 per cent (Pathak *et al.*, 2013). Methane emission can also be reduced by incorporation of organic manure in alternating seasons. The management of crop residues is one of the key components of conservation agriculture, which can potentially reduce GHG emissions.

B. Elevation Potential of Rainfed Ca-based Cropping System's

CA practices have potential to mitigate GHG emissions in rainfed production systems with increased in crop productivity and quality. By adopting zero tillage (ZT) in pigeon pea-castor based cropping system in rainfed situation reduced 24 per cent GHG emissions and significantly cut down the energy input as compared to the conventional tillage. The increase in GHG emissions and energy input were mainly due to increase in the consumption of diesel because of increased number of tillage practices which enhance soil CO_2 emissions and CO_2equivalent emissions (Pratibha *et al.*, 2017). That is why the adoption of conservation agriculture practices (ZT and crop residue retention) in rainfed as well as semiarid regions has huge potential to mitigate GWP because these regions contribute to 58 per cent agricultural production and occupy about 80 per cent of area globally. Therefore, there is a significant potential to reduce emissions of GHG in rainfed ecosystems.

Elevation through Water Management Techniques

The lowering of agricultural CH_4 emissions will largely be achieved through a transition from anaerobic rice production to aerobic, alternate wetting and drying, bed planting of rice, and micro-irrigation. Implementing improved management practices such as adequate mid-season drainage in the field has reported 70-80 percentage of CH_4 emissions reduction compared 4 with continuous flooded rice. GHG emissions can be reduced by raised bed planting of crops which is also an important water management strategy.

Carbon Sequestration in Farming Systems

Farming is known to modify the C cycle, and crop management practices determine the amount of CO_2 emissions into the atmosphere as well as the potential soil carbon sequestration. Effective management of crop residues contributes to enhancement in soil fertility, improvement in soil structure, sequestration of soil carbon, and mitigation of GHG. Tillage and crop management influence the SOC mainly in the 0-60 cm layer. In zero till with residue retention, the SOC content was

118 x 10 kg C/ha/year 3 as compared to 40 × 10 kg C/ha/year in zero till with residue removal. Agroforestry systems can play an important role in sequestering carbon in the above ground and below ground biomass and soil. Improved land and water management practices can be an important strategy for achieving a quantum jump in production per capita while increasing environmental quality and reducing GHG emissions.

Table 6.1: Synergy and Trade-Offs Among the Climate Change Acclimatization and Elevation Measures for South Asian Countries

Management Practices	*Adaptation Process*	*GHG Elevation*	*Synergy or Trade Off*
1.Tillage and crop residue management	The production of zero tillage wheat needs less irrigation and thus, helps to adopt to the limited conditions of water availability. Residue retention under no-till situations helps adapt to excess rainfall and moisture conditions. The cultivation of DSR NOT only saves the irrigation water but also helps to grow crops under limited moisture conditions.	ZT based wheat decreases GHG emissions by 1.5 mg $C0_2$ eq ha^{-1} season as compared to conventionally grown. Increment in SOC sequestration thus reduced the emissions of GHG. Less ground water pumping due to reduction in water application which cut down the energy used for irrigation as well as reduce GHG.	Synergy
2.Water management	In rapidly decling ground water regions, laser land levelling helps adapt to limited water supply conditions.Adapt rainfall variability by expansion of irrigation.	GHG emission may increases by expansion of irrigation because it enhances the energy consumption for irrigation.	Trade -offs
3.Agroforestry	Food security can be improved byAdapting to climate risks such as extreme rainfall and by reducing the soil erosion.	Enhanced the below ground and above ground biomass and thus, increased the carbon sequestration.	Synergy
4.Alternate wetting and drying (AWD)	Reduced irrigation water use by 30 per cent and thus, addresses the water scarcity.	The aerobic situations reduced ground water pumping and thus, reduced the energy use and GHG emission.	Synergy
5.Site-specific nutrient management (SSNM)	Improves soil health, and less fertilizer requirement.	Lesser emissions of N_2O	Synergy
6.Crop diversification for water scarcity	Adapt to low moisture situations.	Reduced groundwater pumping and thus, reduced the energy use and GHG emission.	Synergy
7.Improved livestock management	Improved adaptation to climate shocks.	Reduction in emissions of the CH_4	Synergy

Source: Aryal *et al.* (2019).

Acclimatization and Elevation Measures: Trade-Off's and Synergies

The form of technology and its level of adoption affect significantly the trade-offs and synergies of the acclimatization and elevation measures.

Policy Requirements

Due to the efforts of National Agricultural Research System in the country, several climate resilient practices have been developed. These resilient practices have been field demonstrated and their efficacy already proven. There is a need to integrate these resilient practices into the developmental programmes so that these practices can reach a large number of farmers in the country and thus, can impart resilience. As the impacts of climate change are being observed in various parts of the country and the impacts of rainfall variability and extreme events are widespread and felt in several parts of every year, there is an urgent need to integrate these resilient practices into the policy paradigm of the State and Central Governments. Thus, there is an immediate need to streamline the resilient practices into the ongoing developmental programmes. Further, there is a need to sensitize the policy establishment about the impending impacts of climate change and variability, the likely impacts of extreme events in the years to come and the technological options available to minimize their impact, and the need for their integration in to the on-going developmental programmes.

Future Outlook

Climate change is a global problem and efforts are to be made locally keeping in view the global perspective. There is a need to produce more food in the years to come to meet the diversified food requirements of growing population under resource constrained environments. Continuous research efforts are to be made to develop technologies with less greenhouse gas emissions, which can sequester carbon and utilize the resources efficiently. The developed technologies are to be integrated into the developmental programmes so as to spread them to a large number of farmers in the country to minimize the impacts of variability and to enhance resilience.

Conclusion

- The fact of changing climate and its increasing variability is well established. Under the present situation, the main target is to produce sufficient amount of quality food to meet the demand of growing population while protecting the soil and environment through innovative resource conservation technologies.
- A precise assessment and comprehensive understanding of resource conservation technologies (RCTs) for climate change adaptation the policy establishment about the impending impacts of climate change and variability, the likely impacts of extreme events in the years to come and the technological options available to minimize their impact, and the need for their integration in to the on-going developmental programmes and mitigation options is essential.

- ✰ A wider refinement in the existing technologies and concerted efforts for further research and their adoption by farmers are needed for adaptation and mitigation. The mitigation of greenhouse gases, particularly carbon dioxide emission from agriculture sector, could be attained by increasing carbon sequestration in the soil in the form of soil organic matter.
- ✰ The conservation agriculture (CA) and non-CA based RCTs had imperative role towards climate change resilience. Judicious and efficient nutrient management prompted the use of organic manuring, and biomass recycling will lead to an increase in the soil carbon reserve.
- ✰ The CA-based RCTs with zero or minimum tillage, surface organic residue retention, and legume-based crop rotations have huge potential to augment the use efficiency of natural resources like soil, water, nutrient, carbon, energy, and labour.
- ✰ Besides, diverse methods of cultivation such as mechanical transplanting, the system of rice intensification, direct seeding, in addition to other improved cultivation practices like direct seeding of maize, FIRB system for the establishment of wheat and maize; precise land levelling, precision water management approaches including drip and sprinkler systems and precision agriculture sensor-based reliable nutrient management can rationalise fertiliser application and minimize excessive use of resources.
- ✰ These approaches can advance the sustainability of agro-ecosystem by preserving the natural resource base with higher input use efficiency and also mitigating GHGs emissions.

REFERENCES

Aryal, J.P., Sapkota, T.B., Khurana, R. and Khatri-Chhetri, A. 2019. "Climate change mitigation options among farmers in South Asia". *Environment Development and Sustainability.* PP 1-23.

Behera, B., Das, T.K., Ghosh, S., Kaur, R. and Singh, R. 2018. "Brown manuring in maize (Zea mays): Effects on weed interference and crop productivity". *Indian Journal of Agronomy* 63(4) PP 100- 103.

Bijay, S., Singh, V., Purba, J., Sharma, R.K., Jat, M.L., Yadvinder-Singh, Y., Gupta, R.K., Choudhary, O.P., Chandna, P., Khurana, H.S., Kumar, A., Singh, J., Uppal, R., Vashistha, M. and Gupta, R.K. 2015. Site specific fertilizer nitrogen management in irrigated transplanted rice (*Oryza sativa*) using an optical sensor". *Precision Agriculture*16(4) PP 455-475

Chu, G., Wang, Z., Zhang, H., Liu, L., Yang, J. and Zhang, J. 2015. "Alternate wetting and moderate drying increases rice yield and reduces methane emission in paddy field with wheat straw residue incorporation". *Food and Energy Security*4(3) pp 238-254.

Das, T. K. and Yaduraju, N.T. 2011. "Effects of missing-row sowing supplemented with row spacing and nitrogen on weed competition and growth and yield of wheat". *Crop and Pasture Science* 62(1) pp. 48-57.

Das, T.K. 2008. "Weed science: basics and applications". Jain Brothers, New Delhi

Dixit, A., Khurana, R., Singh, J. and Singh, G. 2017. "Comparative performance of different paddy transplanters developed in India - A revie w". Agri culture Review s 28(4) pp. 262-269.

Edame, G.E., Ekpenyong, A.B., Fonta, W.M. and Duru E.J.C. 2011. "Climate change, food security and agricultural productivity in Africa: Issues and policy directions". *International Journal of Humanities and Social Science* 1(21)

Ghosh, S., Wali, S.Y. and Datta, D. 2017. "Sequential application of herbicides in wheat under Peninsular Indian conditions". *Indian Journal Agronomy* 62(4) pp 535-537.

IPCC. 2019. "Intergovernmental Panel on Climate Change". Climate change and land. https: / /www.ipcc.ch/ srccl.

Kumar, S., Meena, R.S., Lal, R., Yadav, G.S., Mitran, T., Meena, B.L., Dotaniya, M.L. and EL-Sabagh, A. 2018. "Role of legume in soil carbon sequestration". In: Legumes for Soil Health and Sustainable Management, (eds.): Meena, R.S., Das, A., Yadav, G.S. and Lal, R. Springer Nature Singapore: 109- 138.

Lal, R. 2015. "Soil C sequestration in agroecosystems of India". Indian of Journal Society Soil Science 63(2) pp125–143.

Lehman, J., de Silva, J.P., Steiner, C., Nehls, T, Zech, W. and Glaser, B. 2003. "Nutrient availability and leaching in an archaeological Anthrosol and a Ferrasol of the Central Amazon basin: Fertilizer, manure and charcoal amendments". *Plant and Soil*(249)pp 343- 357.

Matocha, J., Schroth, G., Hills, T. and Hole, D. 2012. "Integrating climate change adaptation and mitigation through agroforestry and ecosystem conservation". In: Agroforestry-The future of Land Use, *Advances in Agroforestry* (9) pp 105-126

Meena, R.S., Kumar, S., Yadav, G.S. 2020. "Soil carbon sequestration in crop production. In: Nutrient dynamics for sustainable crop production, (ed.) Nature Singapore: 1-39.

Miao, Q., Rosa, R.D., Shi, H., Paredes, P., Zhu, L., Dai, J., Gonçalves, J.M. and Pereira, L.S. 2016. "Modelling water use, transpiration and soil evaporation of spring wheat–maize and spring wheat–sunflower relay intercropping using the dual crop coefficient approach". *Agri cultural Water Management*(165) pp211-229.

MoEFCC. 2018. Second Biennial Update Report to the United Nations Framework Convention on Climate Change". Ministry of Environment, Forest and Climate Change, Government of India.

Pathak, H., Sankhyan, S., Dubey, D.S., Bhatia, A. and Jain, N. 2013. "Dry direct-seeding of rice for mitigating greenhouse gas emission: Field experimentation and simulation". *Paddy and Water Environment* 11(1) pp 593–601.

Pratibha, G., Srinivas, I., Rao, K.V., Shanker, A.K., Raju, B.M.K., Choudhary, D.K., Srinivas Rao, K., Srinivas Rao, Ch. and Maheshwari, M. 2017. "Net global warming potential and greenhouse gas intensity of conventional and conservation agriculture system in rainfed semi-arid tropics of India". *Atmospheric Environment* (145) pp 239- 250.

Rani, K., Sharma, P., Kumar, S., Wati, L., Kumar, R., Gurjar, D.S., Kumar, D., Kumar, R. 2019. "Legumes for sustainable soil and management". In: Sustainable management of soil and environment, (eds.) Springer Nature Singapore: 193-215.

Richards, M. and Sander, B.O. 2014. "Alternate wetting and drying in irrigated ri ce: Implementation guidance for policymakers and investors". In: Practice brief on climate smart agriculture. Copenhagen, Denmark: CGIAR Research Program on Climate Change, Agriculture and Food Security (CCAFS).

Sharma, A.K., Babu, K.S., Sharma, R.K. and Kumar, K. 2007. "Effect of tillage practices on *Tilletia indica* disease of wheat in rice-wheat rotation of the Indo-Gangetic Plains". *Crop Protection* (26) pp 818-821.

Sidhu, H.S., Jat, M.L., Singh, Y., Sidhu, R.K. Gupta, N., Singh, P., Singh, P., Jat, H.S. and Gerard, B. 2019. "Sub-surface drip fertigation with conservation agriculture in a rice-wheat system: *Agricultural Water Management*(216) pp273-283.

Singh, V.K., Dwivedi, B.S., Singh, S.K., Majumdar, K., Jat, M.L., Mishra, R.P., Rani, M. 2016. "Soil physical properties, yield trends and economics after five years of conservation agriculture-based rice-maize system in N-W India". *Soil and Tillage Res.* (155) pp133-148.

Srinivas Rao, Ch., Girija Veni, V., Prasad, J.V.N.S., Sharma, K.L., Chandrasekhar, C., Rohilla, P.P.,Singh,Y.V.2017. "Improving carbon balance with climate-resilient management practices in tropical agro-ecosystems of Western India"*Carbon Management* 8(2) pp175-190.

Tolossa, T.T., Abebe, F.B. and Girma, A.A. 2020. "Rainwater harvesting technology practices and implication of climate change characteristics in Eastern Ethiopia". *Cogent Food and Agriculture,* 6(1) pp 1724354.

Chapter 7

Role of Smart Villages to Mitigate the Climate Change for Agriculture

Md. Majeed Pasha[1], K.K. Sathian[1] and K.S. Rajashekarappa[2]

[1]Department of Soil and Water Conservation Engineering, KCAET, Thavanur, Kerala
[2]College of Agricultural Engineering, UAS, GKVK, Begaluru, Karnataka

Introduction

Prevailing food insecurity is being worsened by climate change and a growing global population predicted to reach 9 billion by 2050. To meet the needs of this population, food production must increase by at least 70 per cent by 2050 (FAO, 2009). The most challenging part of such an increase is the changing climate particularly in Sub-Saharan Africa, which is known as one of the most vulnerable regions (Ramirez and Thornton 2015).Climate and Agriculture are interrelated processes, both take place on a global scale. Global warming is projected to have significant impact on conditions of affecting agriculture, including temperature, precipitation and glacial run-off. Rising Carbon Dioxide levels would also have effects both detrimental and beneficial, on crop yields. The overall effect of Climate Change on agriculture will depend on the balance of these effects

The climate change is a change in the statistical properties of the climate system when considered over long periods of time. Increasing evidence over the past few decades indicates that significant changes in climate are taking place worldwide as a result of enhanced human activities. Climate change is caused primarily by increases in greenhouse gases such as Carbon Dioxide (CO_2). The presence of the gas in the atmosphere keeps some of the radiant energy received by the earth from being returned to space, thus producing the so called greenhouse effect. These activities accelerated the processes of climate change and increased the mean global temperatures by 0.6°C during the past 100 years, a phenomenon known as global warming. Climate change also effects on food production of India it is sensitive to climate changes such as variability in monsoon rainfall and temperature changes within a season. Every 1°C rise in temperature reduces wheat production by 4-5 Million Tonnes. Small changes in temperature and rainfall have significant effects on the quality of agriculture, horticulture, plantations and medicinal plants. The climate change is something that cannot be prevented but can be reduced by changing the living style of the population and adopting greener technologies. The greener technology includes afforestation, reforestation, using renewable energy sources such as solar energy and wind energy. Adaptation strategies can help to minimize negative impacts to some extent where as mitigation options can help in long run, but There are 'n' number of solutions need to be standardized in the coming years, these need research and policy support.

Impacts of Climate Change on Agriculture in India

1. Climate change can affect crop yields, both positively and negatively, as well as the types of crops that can be grown in certain areas.
2. Climatic changes will affect agriculture through direct and indirect effects on crops, soils, livestock and pests.
3. Increase in atmospheric CO_2 promotes the growth and productivity of C_3 photosynthetic pathway crops.
4. Increase in temperature can reduce crop duration, increase crop respiration rates, alter photosynthate partitioning to economic products, effect the survival and distribution of insect-pest and diseases, hasten nutrient mineralization in soils, decrease fertilizer use efficiencies and increase evapotranspiration.
5. Change of land use, availability of irrigation, frequency and intensity of inter and intra-seasonal droughts and floods, and availability of energy.

Rice	Irrigated rice yield to reduce by 4 in 2020 (2010- 2039) Rainfed rice yield to reduce by 6 in 2020 (2010- 2039)
Wheat	1°C rise in temp reduce yield by 6 mt Yield reduction (6-23 per cent) by 2050
Apple	Lack of sufficient chilling hours, extreme weather events, poor pollination, shifting of apple cultivation from low to high altitude (up to 30 per cent yield reduction)
Cattle and buffalo	Estimated yield loss in milk due to heat stress at 1 8 m t year, which is about 2 of the total production in the country Decline in availability of water may further effect animal productivity
Marine fisheries	Sea surface temperature has increased by 0 2 to 0 3°C along the Indian coast in the last 45 years, and is projected to increase by 2.0 to 3.5°C by 2099 Latitudinal extension in abundance, spawning, breeding activity of several species to be affected

Climate Smart Villages

Climate Smart Villages are areas identified as high risk from the effects of climate change and where local research partners are already conducting work. Prior to the project start-up, a Steering Group made up by researchers and village representatives identify 'climate smart' interventions best suited for that village. The list of activities is introduced to farmer groups and village officials, and the different technologies are prioritized based on yield, resilience potential and other features they find useful. The process is participatory and inclusive as possible, especially encouraging women and more vulnerable groups to participate.

Steps Involved in Setting up a Climate Smart Villages

Selecting the Site

The location of a Climate Smart Village is selected based on its climate risk profile, alternate land-use options, and on the willingness of farmers and local government to participate in the village development activities.

Working with Communities

Community involvement is integral to the success of a Climate Smart Village. Works with existing community groups, consisting of farmers, researchers, rural agro advisory service providers and village officials. They are briefed on the objectives of Climate Smart Villages and encouraged to formally register with the government (if they have not already) to benefit from subsidies on government schemes.

Conducting the Baseline Survey

Researchers conduct a comprehensive baseline study to capture the current socio-economic situation, resource availability, average production and income and risk management approaches of village households. This enables an assessment of the impact of the interventions after a certain period of time.

Prioritizing Interventions

Stakeholders convene to prioritize and test which climate smart technologies and approaches are best suited to their local conditions. Focus group discussions involve farmers in a choice experiment using dummy money to indicate which actions they would most willingly carry out.

Building Capacity

To promote the community's involvement and motivate farmers, a range of tools and approaches are sometimes offered up front. These include rain gauges, improved seed varieties, new livestock breeds, tree seedlings, simple machinery such as zero-till machines, subsidies on index-based insurance premiums and discounts on cell phone SIM cards. Scientists, private sector representatives, and local government organize regular training sessions for farmers on good agricultural practices. At some sites, a small farm is used by the researchers to demonstrate the complete portfolio of interventions.

Monitoring and Evaluating Progress

The lead partner in the village appoints a site coordinator and assistant to provide technical inputs and liaise with institute resource persons. Participating farmers maintain a daily diary of their farm activities and work with the site coordinator to monitor and evaluate the progress of their chosen interventions. These results are digitized and analysed by researchers at the end of every crop season.

Disseminating Outcomes

To spread the message of climate smart agriculture, participatory videos on success stories and testimonials from the pilot villages are screened in nearby villages. Success stories are also widely publicized through local, national and international media. Local government partners organize regular 'farmer field days' to motivate farmers, address their questions and improve on existing strategies.

Components of Climate Smart Villages

Weather

For farmers, information of near-term weather events go a long way in planning climate resilient agricultural production. Farmers are linked to weather information and value-added agro-advisories through radio shows, television, newspapers and mobile phone voice messages. Farmers can use index based insurance schemes to cover risks associated with changes in rainfall and temperature at the different stages of crop growth.

Water

Resilient water management practices which aim at enhancing the efficiency and productivity of water, are critical climate smart interventions. These could include aquifer recharge, rainwater harvesting, community management of water, laser land levelling, water conservation, drip irrigation and on farm water management practices. For instance, drainage is important in places that are prone to becoming waterlogged. Vertical drainage systems are being evaluated in Bihar with the expectation that it will help floodwater seep more quickly back into the natural aquifer, providing the dual benefit of recharge and protecting standing crops.

Carbon

Carbon content in the soil can be increased through agricultural practices such as agroforestry, livestock and manure management, conservation tillage, diversified land-use systems and residue management.

Nitrogen

In Climate Smart Villages, farmers use leaf-colour charts, handheld crop sensors, and nutrient decision-maker tools to decide the most appropriate dosage of nitrogen fertilizers for their crops. This saves on costs and also cuts down on greenhouse gas (GHG) emissions.

Energy

Fuel-efficient agro-machineries, residue management and reduced tillage as interventions to conserve energy and reduce GHG emissions. In some cases, biogas systems are promoted using manure slurry from intensive dairy enterprises

Knowledge

CCAFS and partner organizations arrange cross-site visits of farmers to analogue sites and to other areas practicing climate smart agriculture. An innovative approach of crowd sourcing seeds is used in Climate Smart Villages. A large number of farmers are provided with seed packets of adapted varieties to evaluate those best suited to their local conditions. They provide feedback to researchers to help them develop better varieties.

REFERENCES

Aggarwal P, Zougmoré R and Kinyangi J. 2013. Climate Smart Villages: A community approach to sustainable agricultural development. Copenhagen, Denmark: CGIAR Research Program on Climate Change, Agriculture and Food Security (CCAFS).

CCAFS Working Paper no. 119. CGIAR Research Programme on Climate Change, Agriculture and Food Security (CCAFS). Copenhagen, Denmark.

IPCC – Intergovernmental Panel on Climate Change 2007a: Summary for Policymakers. In: Solomon, S.; Qin, D.; Manning, M.; Chen, Z.; Marquis, M.; Averyt, K.B.; Tignor, M.; Miller, H.L. 2007: Climate Change 2007: The Physical Science Basis. Contribution of Working Group I to the Fourth Assessment Report of the Intergovernmental Panel on Climate Change.

Ramirez-Villegas J, Thornton PK. 2015. Climate change impacts on African crop production.

Chapter 8

Climate Change Effect on Rice Productivity

*Ekta Pandey[1] * and Medha Pandey[2]*

[1]Department of Agricultural Statistics and Department of Agronomy, A.N.D.U.A.&T., Kumarganj, Ayodhya – 224 229, Uttar Pradesh
[2]Integral University Lucknow – 226 026, , Uttar Pradesh

This book addresses aspects of rice production in rice-growing areas of the world including origin, history, and role in global food security, cropping systems, management practices, production systems, cultivars, as well as fertilizer and pest management. As one of the three most important grain crops that help to fulfill food needs all across the globe, rice plays a key role in the current and future food security of the world. Further, this text highlights harvesting, threshing, processing, yields and rice products and future research needs. Supplemented with illustrations and tables, this text is essential for students taking courses in agronomy and production systems as well as for agricultural advisers, county agents, extension specialists, and professionals throughout the industry.

Introduction

Rice has been a staple food source for many cultures globally and is particularly critical for developing countries where other food resources are limited. Rice grains feed more than 3.5 billion people. Rice germinates under a wide range of temperatures, from 10 to 40°C, and requires about 3–6 months to mature. There are two primary methods of rice production: transplanting of rice seedlings from a nursery to paddy rice fields to ensure uniformity and weed control and sowing seeds in a rice field directly to conserve labor. Rice production was particularly enhanced during the green revolution through the deployment of high-yielding uniform rice varieties, especially semi dwarf rice. However, the wide deployment of well-adapted uniform varieties consequently made rice more vulnerable to both biotic and a biotic stressors. Many major rice-growing countries today including

China, India, Japan, Korea, Brazil, the USA, the Philippines, and Thailand have been experiencing old and new obstacles that inhibit the stable production of rice. As such, the question of whether the current methods of rice cultivation are sufficient, or whether we are there yet, or have platitude, remains unanswered.

Importance of Rice

Rice provides 21 per cent of global human per capita energy and 15 per cent of per capita protein. Although rice protein ranks high in nutritional quality among cereals, protein content is modest. Rice also provides minerals, vitamins, and fiber, although all constituents except carbohydrates are reduced by milling. Human Development Report for 1997, approximately 70 per cent of the world's 1.3 billion poor people live in Asia, where rice is the staple food. To some extent, this reflects Asia's large population, but even in relative terms malnutrition appears to affect a substantially larger share of the population in South Asia than in Africa.

The Effect of Rice on the Global Economy

Rice is also the most important crop to millions of small farmers who grow it on millions of hectares throughout the region, and to the many landless workers who derive income from working on these farms. In the future, it is imperative that rice production continue to grow at least as rapidly as the population, if not faster. Rapid population growth puts increasing pressure on the already strained food-producing resources. The aggregate population of the less developed countries grew from 2.3 billion in 1965 to 4.4 billion in 1995. Asia accounted for 60 per cent of the global population, about 92 per cent of the world's rice production, and 90 per cent of global rice consumption. The world's annual rough rice production, however, will have to increase markedly over the next 30 years to keep up with population growth and income-induced demand for food.

Challenges in Rice Production

Major challenges and opportunities for rice grains are embedded during rice propagation, protection, postharvest pest management, processing, and marketing. First, rice diseases have had a major impact on the stability of rice production. For example, the famine in Bengal, India, in 1942 was mainly due to rice brown spot disease. and a rice blast epidemic in Korea caused a major food shortage in the 1970s. The most common rice diseases are caused by fungal and bacterial pathogens. Fungal diseases include rice blast, sheath blight, false smut, rice brown spot, rice stack burn, leaf smut, and grain discoloration diseases include rice blast, bacterial panicle blight, false smut, and kernel smut. Most rice pathogens and insects adapt to their food sources through genetic changes, resulting in more virulent races. Consequently, rice pests are becoming more resistant to pesticides and newly deployed resistance genes.

Suggestion for Production of Rice

Rice thrives in a wide range of geographic and climatic regions, especially where many other crops would fail. As a semi aquatic crop, rice uniquely requires a lot of water as evident by its Chinese name, "water grain." A lack of sufficient clean water is problematic for producing healthy and safe rice as the presence of water is particularly critical during the rice vegetative and reproductive growth stages. There are numerous reasons for the current limit in clean water supply for rice production: depletion of groundwater, increased water usage by human consumption, and contamination due to improper disposal of industrial waste and pesticide contamination. There are numerous reasons for the current limit in clean water supply for rice production: depletion of groundwater, increased water usage by human consumption, and contamination due to improper disposal of industrial waste and pesticide contamination. It is evident that human life in the twenty-first century is much better than that of past centuries due to continued economic growth. As more people are being liberated from starvation, a demand for improved infrastructure to accommodate the incoming population has increased. Therefore, rice is not a competitive crop compared to many other non-essential agricultural products that generate higher income. Rice investment, as such, is among the lowest in many parts of the world.

Conclusion

We believe that scientific and technological innovation will continue to play an important role in increasing rice yield despite changes in socioeconomic and physical environments and problems related to rice production. To meet and sustain the needs of the growing population, the rice production has to be enhanced with good management practices with shrinking availability of land and water resources condition. At present more than 50 per cent of the rice produced in Belize is processed at the Toledo Rice Mill. The mill is obsolete and does not have the equipment required to separate head rice from broken more than a quarter grain. The staff at this mill appears to do a remarkable job, given the operational constraints imposed by inadequate equipment. A major upgrading will be required to allow production of export-quality rice. An extension program to educate maple rice producers about post-harvest moisture and quality control would be beneficial. Other recommendations would be to seek aid from the government in providing subsidies for farmers, educate farmers of the importance of farming and how to go about applying it or even seek funding and grants from other agricultural organizations.

REFERENCES

Global warming, rice production, and water use in China: Developing a probabilistic assessment. Agricultural and Forest Meteorology 148: 94–110.

Projection of effects of climate change on rice yield and keys to reduce its uncertainties. Crop, Environment Bioinformatics 7: 260–268.

Tropical Climate and Its Influence on Rice. IRRI Research Paper Series 20. Los Baños, Philippines, IRRI. Yoshida, S. 1981.

https: //www.slideshare.net/DEVENDRAPALSINGH2/book-chapter-climate-change-and-rice

https: / /www.intechopen.com/books/protecting-rice-grains-in-the-post-genomic-era/introductory-chapter-protecting-rice-grains-in-the-post-genomic-era-are-we-there-yet-

https: / /www.springer.com/gp/book/9783319475141

https: / /sites.google.com/site/bigfallsricemillmisproject/research-page/conclusion

Chapter 9

Effect of Climate Change on Availability of Phosphorus on Soil

Ragini Kumari[1], Mona Kumari[2], Sunita Kumari[3], Geeta Kumari[4], Binod Kumar Vimal[1], Amit Kumar Pandey[5] and Ashutosh Kumar Singh[5]

[1]Assistant Professor, [2]M.Sc. Student
Department of SSAC, BAU, Sabour, Bihar
[3]SMS (Agronomy), KVK, Vaishali, Dr. RPCAU, Pusa, Bihar
[4]Assistant Professor, Department of Microbiology, FBS and H, Dr. RPCAU, Pusa, Bihar
[5]Assistant Professor, Department of SSAC, MBAC, Agwanpur (Saharsa), Bihar

Climate change has become a global concern over the last few decades. Adverse impact of climate change leads to global warming which ultimately results in increase of carbon dioxide (CO_2) concentration of earth's atmosphere which has the potential of increasing greenhouse effect, temperature and variation in precipitation (Touhami *et al.*, 2020). Availability of nutrients is greatly influenced by various environmental and soil factors. Phosphorus (P) is one of the essential nutrient elements found to be affected with the changing climate. Increased CO_2 concentration enhance photosynthetic rate in plant and subsequent growth responses. P acquisition alters under changing climate through changes in root morphology and also by changing in quality and composition of root exudates. Alterations in microbial activity under this changing scenario bring changes in soil P concentration and its transformation (Jin *et al.*, 2015). EC inducted growth stimulation has to be sustained by sufficient nutrient supply to optimize carbon fertilization effect on crop productivity (Kumar *et al.*, 2012). Drought condition resulted in decrease in P availability by affecting environmental factors such as nutrient shortage (Sardans and Penuelas, 2004). To sustain crop productivity in this changing climate scenario, there is requirement of P efficient crop cultivars and systematic nutrient management (Kumar *et al.*, 2011).

Introduction

Climate change has significant impact on agricultural production system. The atmospheric carbon dioxide concentration has been steadily increased during the last 12 years at the rate of 1.9µmol/mol/year by the middle of the 21st century. As a consequence earth's average surface temperature is expected to increase 2-4.5 °C which can affect the soil processes including the nutrient availability and its uptake by crops. Increase temperature further decreases soil water availability which also results in a decrease in nutrient uptake by roots. Phosphorus is one of the essential nutrient elements affected with the change in climatic factors as well as variations in soil conditions. In this changing climate scenario, there are significant changes in various morphological and biochemical properties in soil and plant which ultimately changes the soil P concentration and its transformation process.

Factors Affecting Phosphorus Availability

Carbon Dioxide (Increased)

- Enhance photosynthetic rate
- Alters P acquisition through changes in root morphology and rooting depth.
- Quantity and quality of root exudates.
- Microbial activity in rhizosphere.

Water and Heat Stress

- Dry matter yields of all plant decreases under water and heat stress which ultimately can limit plant growth.
- Elevated temperature decreases soil moisture content that can affect soil phosphorus uptake by changing the P flux to the root surface as well as growth and physiology of root system.

Mechanism by which Elevated CO_2 Impacts Plant Phosphorus Nutrition

Jin *et al.* (2015) proposed various mechanisms by which elevated CO_2 concentration affects plant phosphorus nutrition. Elevated atmospheric CO_2 concentration can enhance photosynthetic rate in plants which can act as a carbon fertilizer which ultimately stimulate growth of many plant species. An increase in growth of plant will need an increased supply of essential plant nutrient such as 'P' which plays an important role in plant metabolism. Under elevated CO_2, the growth rate of shoot increases together with an increase in the carbon allocation to root and increase in root carbon will stimulate root growth and bring change in root morphology due to which there is increase in the spread of root through the root zone which would lead to increase in P uptake. Exudates secretion also increases from root in response to carbon allocation to roots. These exudates like organic acid and carboxylates mobilize sparingly soluble P via chelation and ligand exchange and also stimulate secretion of phosphatase to mineralize organic P forms in soil.

Soil pH also changes in response to elevated CO_2 concentration. Change in CO_2 concentration bring changes in quality of organic acid anions and associated proton released in exudates from plant roots lead to pH change in rhizosphere. Also there are increased activities of rhizosphere microorganism under elevated CO_2 which can increase CO_2 concentration in soil and this CO_2 dissolve in soil water to form carbonic acid and as a result pH in rhizosphere decreases which can influence the solubility of P in soil and its further uptake. There is also increase in C/P ratio under elevated CO_2 because there is decreasement in P concentration due to increase in microbial biomass under high CO_2 concentration which will compete for more phosphorus resulting in P immobilization. Elevated CO_2 is also able to specifically affect the abundance of some microbial genera which may directly facilitate P solubilization in rhizosphere. As CO_2 increases carbon allocation to roots, Arbuscular mycorrhizal fungi (AMF) hyphal network is also enlarged resulting in nutrient absorption being significantly increased.

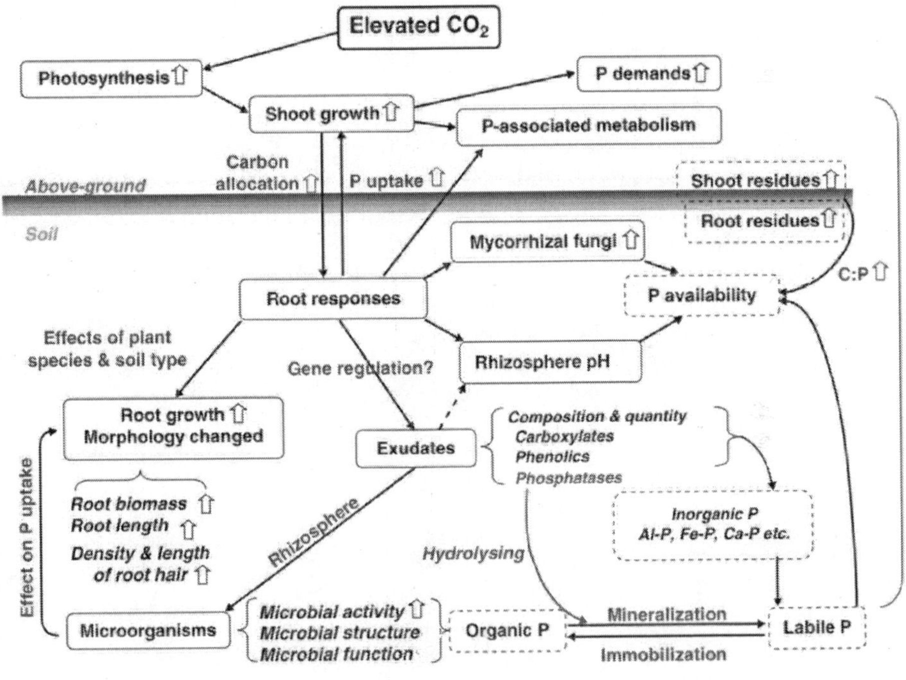

Figure 9.1: Phosphorus Nutrition is Regulated under Elevated CO_2 Condition (Jin *et al.*, 2015).

Effect of Elevated CO_2 on Soil pH, Soil Total Carbon and Olsen P

1. Touhami *et al.* (2020) found that soil pH was not affected by eCO_2 because soil was already near the optimum pH for P availability (Figure 9.2). Hence, acidification would likely not increase P availability. But in neutral to alkaline pH, available P increases in cropping system after eCO_2 due

to acidification of soil by the release of various exudates such as organic acid *etc.*

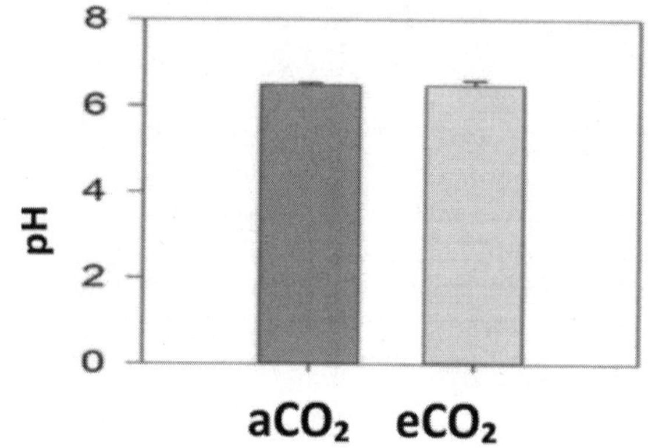

Figure 9.2: Impact of eCO_2on Soil pH.

2. Touhami *et al.* (2020) found that total soil carbon was increased significantly under eCO_2 in comparison to ambient CO_2 condition because under exposure to elevated CO_2there is increase in rhizodeposition, root turnover and plant residue. These increased carbon input is responsible for increase in total carbon (Figure 9.3).

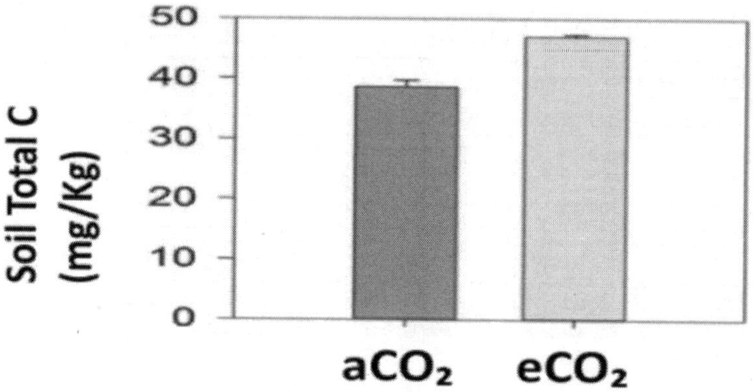

Figure 9.3: Impact of eCO_2on Soil Total Carbon.

In this graph, Olsen P was lower under eCO_2 in comparison to aCO_2treatment because Olsen P fraction is readily available for plant and microbial uptake. As growth rate of plant increases, so it will need more supply of P which will be fulfill by this fraction (Figure 9.4).

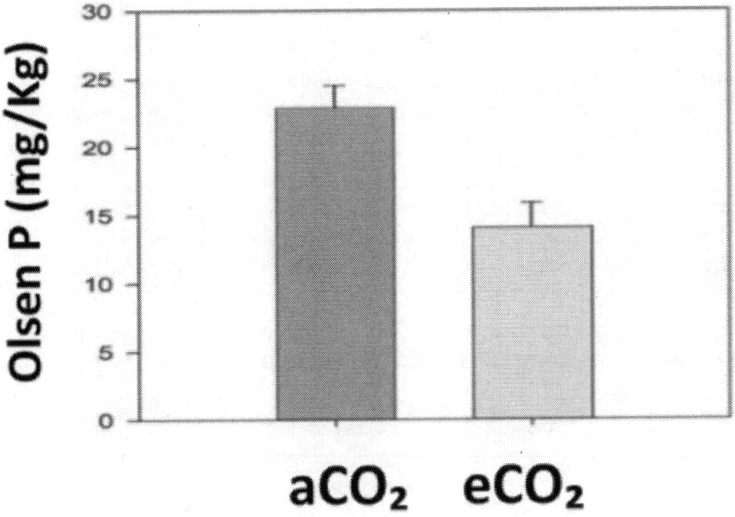

Figure 9.4: Impact of eCO_2 on Soil Olsen P.

3. Impact of elevated CO_2 on Acid phosphatase activity and Alkaline phosphatase activity

Touhami *et al.* (2020) reported that there is increase in alkaline phosphatase activity under elevated CO_2 condition compared to ambient condition because phosphatase enzyme are largely derived from microorganisms in the soils and due to increase of microorganisms in response to elevated carbon dioxide, alkaline phosphatase activity also subsequently increased.

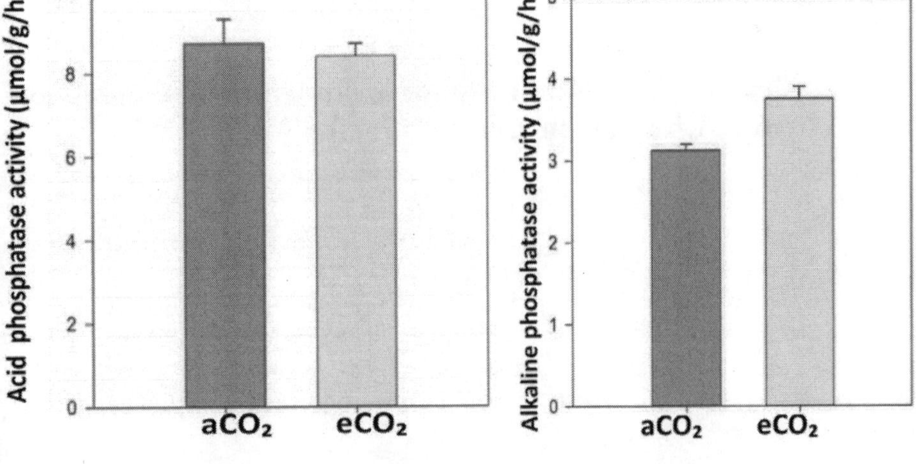

Figure 9.5: Impact of Elevated CO_2 on Acid Phosphatase Activity and Alkaline Phosphatase Activity.

Effect of atmospheric CO_2 concentration, temperature and P-levels on dry weight, P concentration and P uptake of whole plant

Kumar *et al.* (2012) studied that EC induced increased in dry weight of whole plant is due to carbon fertilization effect whereas P uptake, per cent P derived from fertilizer, Fertilizer P uptake increased under eCO_2 because there is better root and shoot growth under elevated CO_2 which leads to higher P requirement. Under elevated temperature condition all the parameters decreases due to thermal stress. Under different phosphorus levels all parameters were highest under 200 per cent P treatment (Table 9.1).

Table 9.1: Effect of Atmospheric CO_2 Concentration, Temperature and P-levels on Dry Weight, P Concentration and P Uptake of Whole Plant

Parameters	Carbon dioxide levels		Temperature levels		Phosphorus levels		
	ambient	elevated	ambient	elevated +3 °C	control P	100% P	200% P
Dry weight (g/ pot)	10.5	13.9	14.3	10.2	10.4	12.0	13.5
P uptake (mg/pot)	17.1	27.0	25.3	18.7	14.0	23.0	29.0
% Pdff	50.9	55.9	54.2	54.0	-	42.2	48.6
Fertilizer P-uptake (mg/ pot)	8.70	15.1	13.7	10.1	-	9.71	14.1

Source: Kumar *et al.* (2012).

Effect of Elevated CO_2 and Phosphorus Nutrition on Growth and Morphological Changes in Cereal Species

Pandey *et al.* (2014) reported that in all the three cereal species, plant grown with sufficient P produced more than 3 leaves compared to those growths with low P condition. PDW-233 wheat variety and WSP-540-2 rye variety also produced tiller in response to eCO_2 with sufficient P. This is may be due to carbon dioxide enrichment which had positive effect on growth of plant under sufficient P. Root growth were more in all cereal in response to elevated CO_2 grown at low P as compared to aCO_2 because under restricted P supply root growth is enhanced so that larger root system can easily capable of exploring greater soil volume and can capable of sufficient P uptake for plants (Figure 9.6).

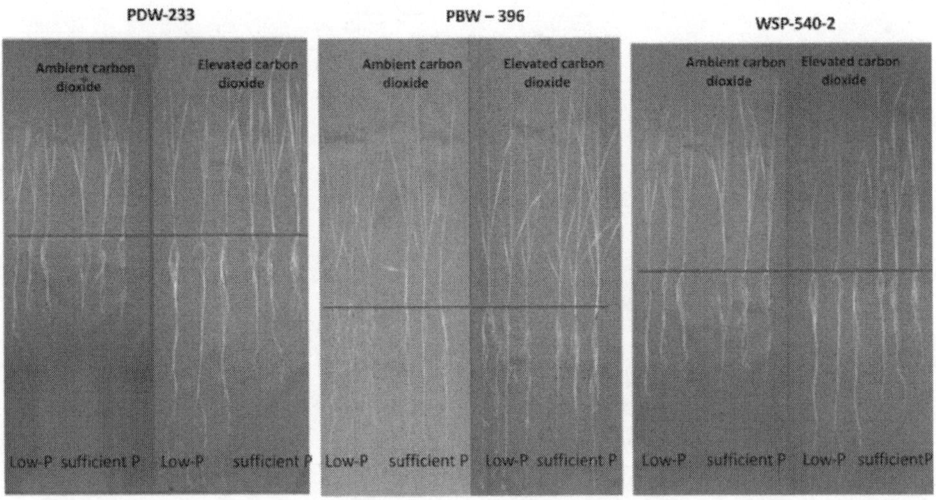

Figyre 9.6: Effect of Elevated CO_2 and Phosphorus Nutrition on Growth and Morphological Changes in Cereal Species (Pandey *et al.*, 2014).

Effect of different CO_2 and Phosphorus Levels on Shoot and Root Dry Weight, Root Length and Root : Shoot Ratio of Cereal Varieties

Pandey *et al.* (2014) studied the effect of different CO_2 and phosphorus levels on shoot and root dry weight of cereal varieties by using different treatment combinations *i.e.*eCO_2_elevated CO_2 level (700µmol/mol), aCO_2 – ambient CO_2 level (380µmol/mol), -P (low P) – 2µM and +P (Sufficient P) - 500µM. They found that variation in shoot and root dry weight, root length and root: shoot ratio of cereal varieties. Shoot dry weight averaged over all cereal species were significantly higher in plant raised with sufficient P under elevated CO_2 condition compared to ambient CO_2 concentration as carbon fertilization had positive effect on dry matter accumulation under sufficient - P. Plant grown at low P shows a significant increase in root dry matter at eCO_2 in comparison to ambient CO_2 because root growth is enhanced to fulfill its nutritional requirement by exploring larger soil volume. Root length averaged over CO_2 and P level was higher in PDW-233. All cereal species recorded higher root to shoot ratio under low P irrespective of CO_2 level. Reason for this is that under restricted P-supply, increased root growth is advantageous as it allows plants to obtain and utilize the available mineral supply efficiently by exploring more soil volume by larger root system (Figure 9.7).

Impact of CO_2 Concentration on P-pools in Living Tree Biomass

Khan *et al.* (2010) reported that P concentration in total living biomass constituting of stems, leaves, fine and coarse roots was non-significantly increased under elevated CO_2 (FACE-Free air carbon dioxide enrichment) on an average

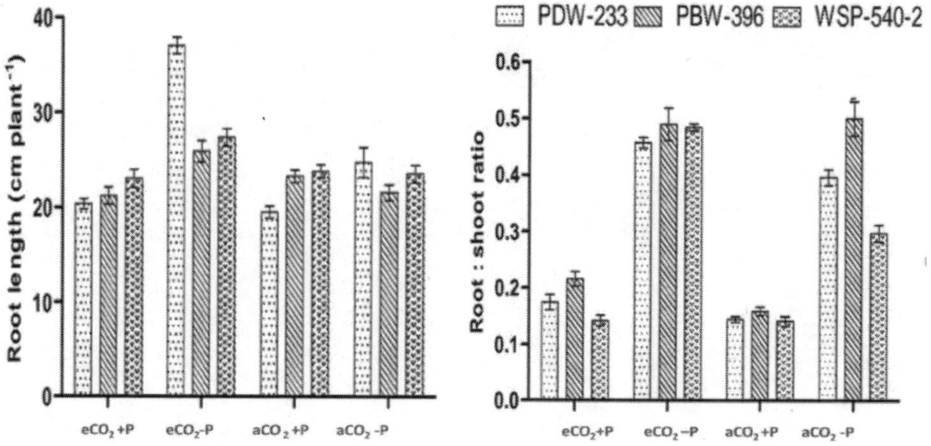

Figure 9.7: Effect of different CO_2 and Phosphorus Levels on Root Length and Root : Shoot Ratio of Cereal Varieties (Pandey *et al.*, 2014).

across species. When looking at separate biomass pools, the amount of P held in root increased significantly on an average across species. Reason for this increase in P concentration in tree living biomass is that as elevated CO_2 lead to larger tree biomass due to which there is increase demand of P and ultimately P concentration increases in living biomass (Figure 9.8).

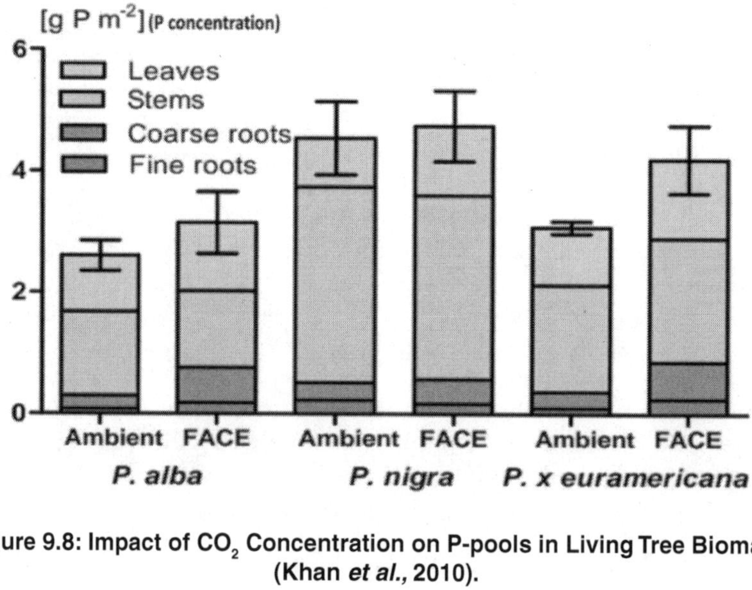

Figure 9.8: Impact of CO_2 Concentration on P-pools in Living Tree Biomass (Khan *et al.*, 2010).

Effect of different CO_2 Concentration and Temperature Conditions on Soil Microbial Biomass P Organic Content in Root Exudates and Phosphatase Enzyme Activities under Tropical Flooded Rice Soil

Bhattacharyya *et al.* (2013) worked on effect of different CO_2 concentration and temperature conditions on Soil Microbial Biomass P, organic content in root exudates and Phosphatase enzyme activities under tropical flooded rice soil by using different treatment combination *i.e.*, UC - unchamber control (Open field ambient CO_2), CC – ambient CO_2 in control chamber, EC – chamber with elevated CO_2 and ECT –chamber with elevated CO_2 and elevated temperature. They reported that-

i. Changes in Soil Microbial Biomass P under Tropical Flooded Rice Soil

The microbial biomass P is measure of the living component of soil organic matter. It is directly released from microbial cells when microorganisms die and easily decomposed and becomes readily available to plants. Soil microbial biomass P was significantly increased under EC as well as ECT compared to that of ambient condition at different growth stage of plant. This may be due to higher root exudates content and increased available P in soil (Figure 9.9).

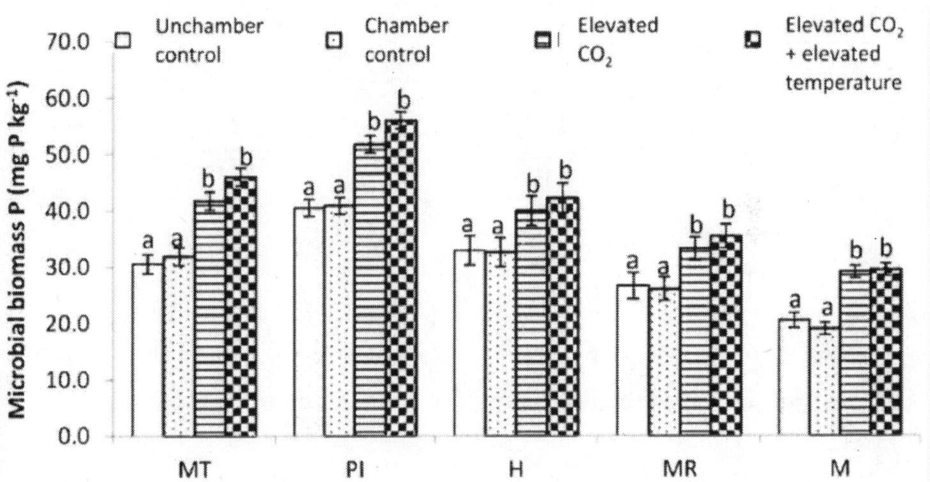

Figure 9.9: Soil Microbial Biomass P under Elevated CO_2 and Temperature in Tropical Flooded Rice Soil.

ii. Changes in Organic Content in Root Exudates under Tropical Flooded Rice Soil

Acid content of all the four acids were significantly higher under EC and ECT compared to that of ambient CO_2 condition. This increasement is because living roots continually release organic carbon into rhizosphere in the form of root exudates.

The quantity and quality of root exudates was affected by increase in plant biomass. These soluble root exudates contain organic acid, polymer degrading enzymes *etc.* which affect nutrient availability by enhancing mineral weathering, nutrient mobilization as well as their uptake. In this way root exudates are beneficial in making phosphorus available (Table 9.2).

Table 9.2: Different Organic Acid Content in Root Exudates under Elevated CO_2 and Temperature in Tropical Flooded Rice Soil

Organic acid (µ mol/plant/day)	Treatments	At maturity of rice
Acetic acid	UC	49
	CC	50
	EC	54
	ECT	57
Malic acid	UC	146
	CC	143
	EC	156
	ECT	157
Tartaric acid	UC	569
	CC	568
	EC	599
	ECT	606
Citric acid	UC	12
	CC	13
	EC	15
	ECT	17

Bhattacharyya *et al.* (2013).

iii. Changes in Phosphatase Enzyme Activities under Tropical Rice Soil

Acid and alkaline phosphatase in soil was strongly influenced by the exposure of elevated CO_2 and temperature throughout the plant growth stage. There was significant increase in both acid and alkaline phosphatase activity under EC as well as ECT condition. An increased phosphatase activity could be consequence of an improved root development or an increased microbial activity. Under elevated condition there is alteration in quantity and quality of root exudates which influence the enzyme activities. As phosphatase is a specific enzyme, that is linked to P mineralization and its availability. Therefore elevated CO_2 indirectly affects P availability (Figure 9.10).

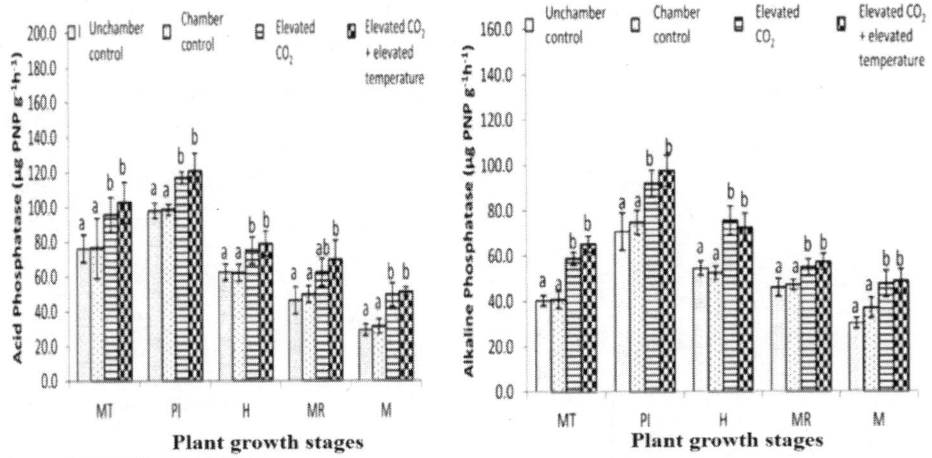

MT – Maximum tillering
PI – Panicle initiation
H – Heading
MR – Mid-ripening
M – Maturity

(Bhattacharyya *et al.*, 2013)

Figure 9.10: Soil Acid Phosphatase Activity and Alkaline Phosphatase Activity under Elevated CO_2 and Temperature in Tropical Flooded Rice Soil

Effect of atmospheric CO_2, Temperature and Phosphorus Levels on Various P Fractions

Kumar *et al.* (2011) reported that organic P in wheat rhizosphere was reduced at EC as well as at ET whereas gradient enrichment was observed with increasing P supply. Reason for decrease is that as organic P is an important source of available P. So, enhanced P demands by plants under EC have caused depletion of the organic P in rhizosphere. Increased activity of phosphatase enzymes under P deficiency induced by EC and ET have resulted into greater mobilization of originally bound P leads to depletion of organic- P. In inorganic P fraction, except residual P remaining all the inorganic P fractions were enriched in rhizosphere of wheat grown under both individually or jointly elevated CO_2 and temperature whereas it increases with increasing supply of phosphorus. Reason for this increase is that a part of applied fertilizer P was not taken by the plant might have contributed to enrichment of these fractions. Enhancement in inorganic P fraction could be due to that a portion of P released from organically bound fraction into soil solution may have contributed to the various inorganic P fractions in the rhizosphere. Residual P, unlike the entire inorganic P fraction declined under EC and ET. Residual P is a very stable P- pool, contribution from this fraction is expected only in exhaustive cropping system with very high demand of plant available P in soil. Under high level of CO_2 and temperature, increased P demands by growing plant have forced the contribution from this very stable P fraction in soil. Total P show a non-significant decrease and increase under EC and ET respectively, while increases were significant with increasing supply of P.

Table 9.3: Effect of Atmospheric CO_2, Temperature and Phosphorus Levels on Various P-Fractions in Wheat Rhizosphere at Maturity

P fractions (mg/kg)	CO_2 levels		Temperature levels(°C)		Phosphorus levels		
	ambient	elevated	ambient	elevated 3°	Control P	100% P	200% P
Loosely bound P	11.6	13.2	10.7	10.3	10.3	12.9	14.0
Al - P	41.8	44.3	39.1	41.9	41.9	42.0	45.3
Fe- P	25.7	27.6	24.0	25.2	25.2	29.1	25.6
Reductant slouble P	65.5	70.4	62.8	68.5	68.5	67.3	68.0
Ca - P	213	223	207	214	214	220	220
Residual - P	47.3	35.0	50.2	41.6	41.6	39.8	42.0
Inorganic - P	405	413	394	401	401	411	415
Total - P	570	563	564	569	552	566	582
Organic - P	165	150	170	151	151	155	166

Kumar *et al.* (2013).

Temporal Change in P Mineralization Rate with a Significant $CO_2 \times$ Time Interaction

Hasegawa *et al.* (2015) showed through this graph, the vertical dotted line indicates the point at which CO_2 treatment began. Although P mineralization rate were lower in eCO_2 condition compared to ambient during the pretreatment period. However there is substantial increase in net P mineralization rate during 1st three months immediately after CO_2 treatments were initiated. The observed increase in P mineralization rate early in this study is due to increased exudation of labile C which enhances microbial decomposition of soil organic matter (priming) leading to increased nutrient availability. Early eCO_2 enhancement of P mineralization did not persist beyond the first six month in this study. Although statistically they are at par but there is some absolute difference between elevated and ambient condition which may be due to effect of moisture, temperature and microbial activity as reported by them (Figure 9.11).

Impact of Drought on Labile P Concentration

Sardans and Penuelas (2004) studied that total labile P was highest in DD samples (run-off + rainfall exclusion) than control sample. Similar results were found for labile organic P whereas the highest labile inorganic P was found in control soil samples. These results indicate that in this condition soil accumulates P in non-directly available forms. The high organic P fraction may be due to high levels of organic matter, due to low mineralization rate under low water availability. Soil phosphatase activity has been decrease in response to drought enhancement (Figure 4.12).

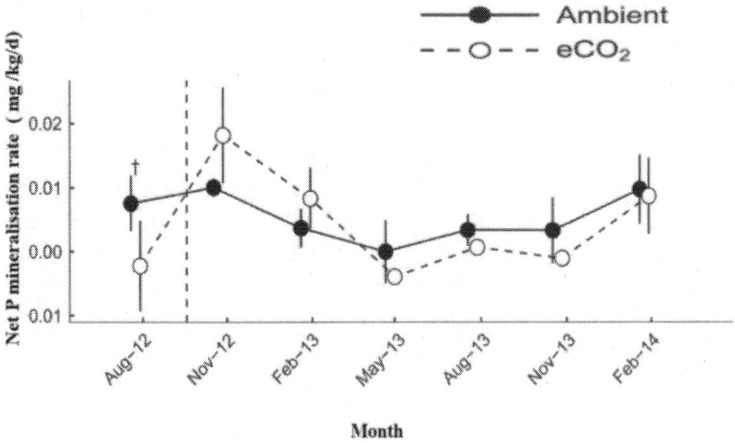

Figure 9.11: Temporal Change in P Mineralization Rate (Hasegawa *et al.*, 2015).

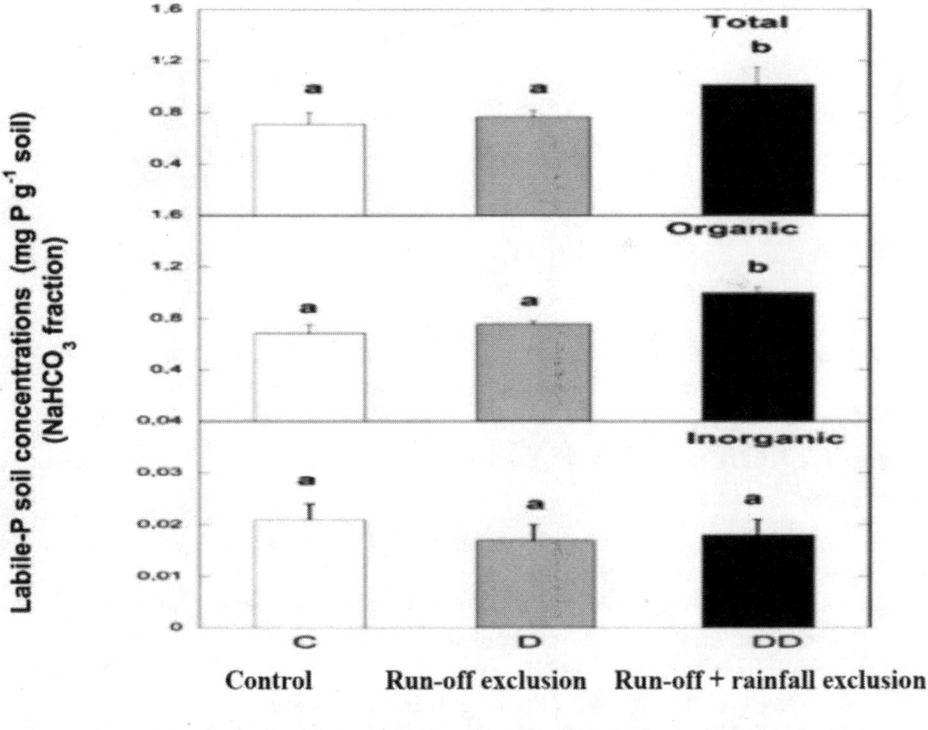

Figure 9.12: Labile-P Soil Concentration Fractions in the Control Plots (C) and in the Treatment Plot D (Affected by run-off exclusion only) and DD (Affected by both run-off and rainfall exclusion) (Sardans and Penuelas, 2004).

Effect of Moisture on Pi/Po Ratio of the Labile Soil P Fractions

Sardans and Penuelas (2004) reported that the Pi/Po ratio is used to evaluate the effects of moisture stress on mineralization capacity of soil. Lower water availability decreased the Pi (mineralized) with regard to the Po (non-mineralized). This effect was marginally significant when comparing soil from control with soil in drought plot affected by run-off exclusion only but clearly significant when comparing control soils with soil in drought plot affected by both run-off and rainfall exclusion. Increase in Po soil fractions in drought plot mainly due to increase in litter fall due to slow microbe mineralization rate at lower water availability (Figure 9.13).

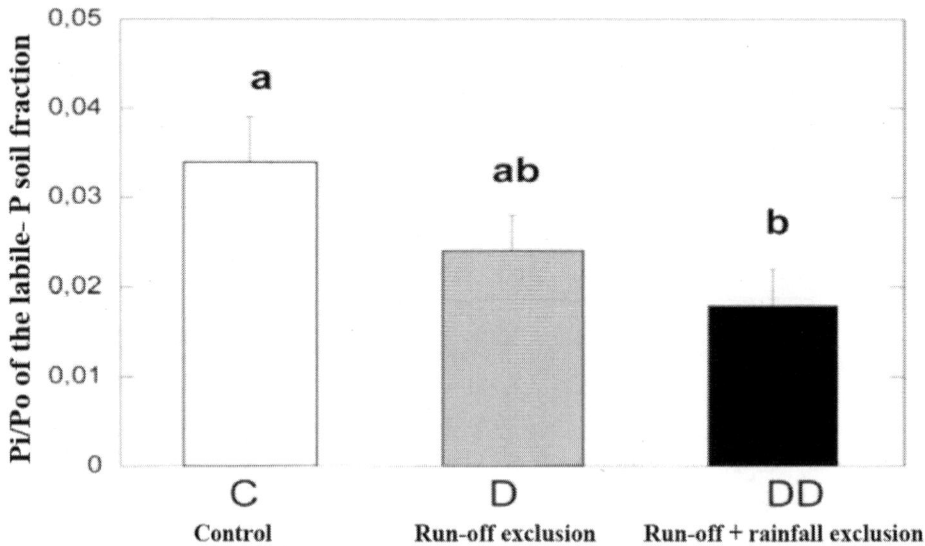

Figure 9.13: Pi/Po Ratio of Labile-P Soil Fraction in the Control Plot C and in Treatment Plot D and DD (Sardans and Penuelas, 2004).

Impact of Drought on P Concentration in the Tree Biomass Fractions

Sardans and Penuelas (2004) P concentration in the different plant fractions in the control plots and in the drought plots were non-significantly different except for those of the fine roots, which were lower in drought plants than in the control plots. Decrease in total plant P content is due to increase in P loss in stand biomass and also due to slower mineralization rate as a consequence of water deficit effect on enzyme activity. Decrease in P concentration was mainly in fine stems, roots due to higher P allocation to photosynthetic structure for avoiding a decrease in photosynthetic capacity caused by a deficit nutrient supply (Figure 9.14).

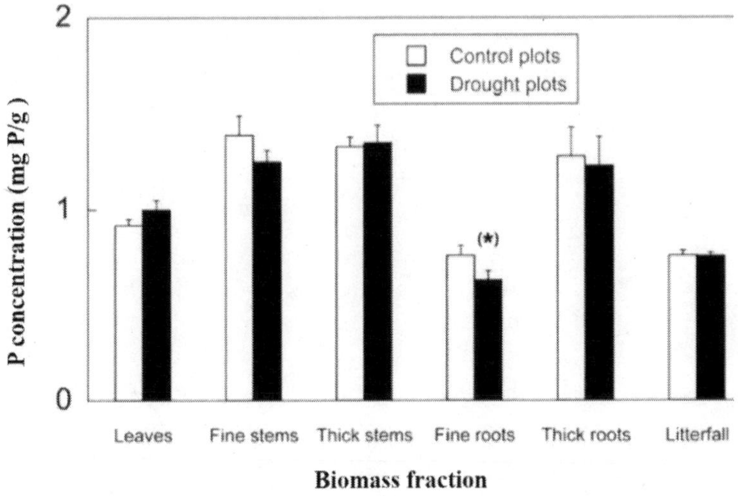

Figure 9.14: P Concentration in Various Tree Biomass Fractions (Sardans and Penuelas, 2004).

Conclusion

Enhanced organic acid in root exudates coupled with higher soil phosphatase activities under elevated CO_2 resulted in increase rate of soil P solubilization leading to higher plant P uptake. EC induced growth stimulation has to be sustained by sufficient nutrient supply to optimize carbon fertilization effect of higher CO_2 on crop productivity. The greater increase in P uptake in elevated CO_2 condition is because average distance of diffusion of nutrient to root surface will decrease due to greater root length density. But there is indirect synergistic effect of increased drought on other environmental stress factors such as nutrient shortage which ultimately decreases P availability.

REFERENCES

Bhattacharyya, P., Roy, K. S., Dash, P. K., Neogi, S., Shahid, Md., Natak, A. K., Raja, R., Karthikeyan, S., Balachandar, D., Rao, K. S. (2013). Effect of elevated carbon dioxide and temperature on phosphorus uptake in tropical flooded rice. *European journal of* agronomy 53,28-37.

Hasegawa, S., MacDonald, C. A., Power S. A. (2016). Elevated carbon dioxide increases soil nitrogen and phosphorus availability in a phosphorus limited Eucalyptus woodland. *Global change Biology* 22,1628-1643.

Jin, J., Tang, C., Sale, P. (2015). The impact of elevated carbon dioxide on the phosphorus nutrition of plants. *Annals of Botany* 116; 987-999.

Khan, F. N., Lukac, M., Miglietta, F., Khalid M., Godbold, D. L. (2010). Tree exposure to elevated carbon dioxide increases availability of soil phosphorus. *Pak. J. Bot,* 42(2), 907-916.

Kumar, M., Swarup, A., Patra, A. K., Chandrakala, J. U., Manjaiah, K. M. (2012). Effect of elevated CO_2 and temperature on phosphorus efficiency on wheat grown in an inceptisol of subtropical India. *Plant soil and Environment,* 58; 230-235.

Kumar, M., Swarup, A., Patrakaar, A. K., Purakayastha, T. J., Manjaiah, K. M., Rakshit, R. (2011). Elevated carbon dioxide and temperature effects on P-dynamics in rhizosphere of wheat grown in a typical haplustept of Sub tropical India. *Agrochimica,* Vol. LV - N. 6; 314-331.

Pandey, P., Dubey, K. K., Ahmad, A., Nilofar, R., Verma, R., Jain, V., Zinta, G., Kumar, V. (2014). Elevated CO_2 improves growth and phosphorus utilization efficiency in cereal species under sub-optimal phosphorus supply. *Journal of plant nutrition.*

Sardans, J., and Penuelas, J. (2004). Increasing drought decreases phosphorus availability in an evergreen Mediterranean forest. *Plant and soil,* 267; 367-377.

Touhami, D., Dowell, R. W. Mc., Condron, L. M., Lieffering, M., Newton, P. C. D. (2020). Long term atmospheric carbon dioxide enrichment decreases soil phosphorus availability in a grazed temperate pasture. *Geoderma*378.

Chapter 10

Mitigation of Climate Change Risk in Aquaculture

Raj Kumar, Akhil Gupta and Sahar Masud

Fisheries Unit, Faculty of Agriculture, Main Campus, SKUAST-J, Chatha

Introduction

As the global population increases, there is no doubt that there will be corresponding increase in demand for food most especially protein. Production from fisheries and aquaculture sector is essential for food security in view of present increase in population. Despite the reliability on fisheries and aquaculture to supply the animal protein needed by the world population, climate change seems to have great potential to significantly reduced production and increase instability in the sector. In order to achieve food security and fisheries development goal, climate smart aquaculture which is an adaptation to climate change will be necessary. The present lecture therefore discusses climate smart aquaculture as veritable approach for increasing fish production in the face of climate change trend.

A number of changes already evident can be attributed to climate change; these mainly involves drastic change in weather condition, reduction in water levels, changes in hydrological regimes of inland water, heavy wind storm, excessive sunshine, increased incidence of flooding and droughts. The effects of these changes have resulted in changes in fish behavior, more chances of fish disease infestation and uncertainty in fish production from inland and aquaculture systems. Climate smart adaptation and mitigation strategies such as use of cover/shades over ponds during dry weathers, by adjusting time of stocking, stocking of fish species and minimum stocking stage (advanced fry, fingerlings, even advanced fingerlings or stunted yearlings) that can better adapted to climate change impacts, will plays an important role to increase the resilience and adaptive capacity of communities and ecosystems.

Other adaptations strategies include digging bore wells to supply water during dry weathers and well-structured drainage system which helps against flooding. Integrated farming system (Fish farming based farming system) is also an important adaptive measure to combat the possible climate change risks of failure especially in case of single crop; which has even great potential for increasing employment and per unit area income. Further, use of low carbon producing energy source in the production and processing of product from aquaculture is a viable means to mitigate against climate change, example include the use of gas or electricity rather than charcoal in fish smoking. Aquaculture wastewater treatment before discharge is also a good mitigation practice been recently developed.

Climate change will have significant impacts on fisheries and aquaculture. Climate smart aquaculture will respond to these changes by boosting adaptive capacity and resilience both of communities and the ecosystems on which they depend. It is important therefore, to ensure adaptation and mitigation in response to climate change so as to safeguard the sustainable fish production and food security improvement. It is now widely accepted that climate change is no longer simply a potential threat, it is unavoidable; a consequence of 200 years of excessive greenhouse gas (GHG) emissions from fossil fuel combustion in energy generation, transport and industry, deforestation and intensive agriculture (IPCC, 2007). Small-scale fisheries and aquaculture have contributed little to the causes of climate change but will be amongst the first sectors to be affected the most. Like Climate change has significant impacts on freshwater and marine aquatic systems and hence has an adverse effect on the country's fisheries and aquaculture. The effect of these changes has resulted in elevated water temperatures (IPCC,2007) which affects fish physiological processes, thereby affecting spawning, survival of the juveniles, recruitment into the exploitable phase of population, population size, production and yield. The impacts of increased flooding of the freshwater bodies has been negative through erosion of watershed, destruction of fish feeding and breeding habitats, decrease in primary productivity and alteration of the normal resilience of the aquatic systems, or positive in expansion of aquatic habitats for primary and fish productions especially during the dry season. Drought incidence draws down the lakes and gives rise to insufficient flow in the river basins which ultimately impacts on spawning and primary production thereby affecting fish production. Rise in the water level of the ocean lead to intrusion of more salty water into the river delta areas and inundation of the coastal low-lying areas, thus affecting distribution of both the freshwater and marine fishes as a result of changes in the physical and chemical properties of the waters.

The impacts ultimately affect fish population, production and supply, thereby affecting the livelihoods of over millions people engaged in the primary and secondary sectors of the fisheries industry, as well as food security. Considering the importance of fisheries sector to the national economy: provision of food supplies, enhancing food security, creation of employment opportunity and income generation, it is important to safeguard the sector from the effect of climate change whose threat have started been felt by small scale fisheries investors and resource poor fisher folks. Innovative policies and investment programmes are needed to help the rural poor respond and adapt to a changing climate, and anticipate, absorb and recover from climate shocks and stresses. Strategic plan is needed to combat the observable and projected impacts of climate change on fisheries and aquaculture in order to protect the aquatic fauna as well as the livelihoods of the fishing communities and food security.

Importance of Fisheries and Aquaculture

Fisheries and aquaculture plays an important role in addressing nutritional and livelihood security, especially of the rural poor in developing countries. It provides the economic and nutritional benefits to the mankind since time immemorial. Fish are rich sources of protein, essential fatty acids, vitamins and minerals. The fats and fatty acids which are present in fish, particularly Omega 3 fatty acids, are highly beneficial and are difficult to obtain from other food sources. The growing gap between supply and demand globally will impact on the health and nutrition of low income families, unless efforts are made to increase the production to meet the growing demand. Aqua farming means culture of commercially important fish species or some other aquatic organism under controlled conditions. Aquaculture is a new name for what once we called 'fish culture. Present concept of fish farming/aquaculture incorporates culture of all aquatic organisms by following certain management techniques which includes water quality, choice food *etc.* and to protect them from unwanted predators, diseases, pollutants or any other things which are harmful to them. The aquatic organisms which are normally used for aquaculture for food purpose include fishes, prawns, shrimps, crabs, mussels and some live food organisms like algae and zooplankton. Aquaculture is essentially an Asian farming practice. India is endowed with vast and varied aquatic resources, of which only a small portion is being utilized for aquaculture practices. Aquaculture continues to increase in volume and value of output in many countries of the world, filling the gap between the supply and demand for fish and fishery products, improving nutrition and contributing to the household economy, particularly in rural areas. There is immense scope for the betterment of mankind through aquaculture. Currently, China leads in Aquaculture production in the world followed by India, but the difference in production is almost 8-9 times. In India, the average growth rate of Aquaculture is about 8 per cent.

Fisheries and Aquaculture in J&K

In Jammu and Kashmir State, there are vast network of water resources in the shape of cold water streams, perennial rivers, lakes, reservoirs, sars, irrigation canals and about 250 high altitude lakes spread over an area of about 40,000 hectares. Besides there are a great number of community and private ponds which plays an important role in the fish production enhancement for state. As per the website of Department of Fisheries J&K Government, total fish production of the state was about 20.70 thousand tons (2017-18). Out of which 482 tons was contributed by trout fish. The hill states of our country are endowed with rich diversity of water resources which comprise of rivers, streams, irrigation canals, natural lakes, man-made reservoirs and various types of wetlands. A total of 258 fish species inhabits coldwater belts of the country out of which some species like mahseer, snow trout and exotic trout are of commercial importance. Since long time, the fish and fisheries of these hilly water resources are providing an important source of nutrients especially protein to the local inhabitants of hill regions. In order to get maximum return per unit area, aquaculture has been identified as one of the most important component of the fisheries.

Brief Idea about the Common Aquaculture Practices

Classification

Aquaculture may be conducted in all the three types of aquatic environments

Freshwater Aquaculture

It involves the culture in the water bodies having salinity level of less than 0.5 parts per thousand (ppt).

Brackish Water Aquaculture

It involves the culture in the water bodies having salinity level ranges from 0.5 to 30 ppt.

Mariculture or Sea Farming

It involves the culture in the water bodies having salinity level of more than 30 ppt. The species of flora and fauna inhabiting the three types of water bodies are accordingly called freshwater species, brackish water species and marine species. Freshwater which is most extensively used sector of aquaculture, is further divided into two segments.

Cold Waters

Cold waters of higher altitudes having temperature range of less than18°C and

Warm Waters

Warm waters of plains having temperature range of more than 18°C. Aquaculture practices in these waters are therefore, called coldwater aquaculture and warm water aquaculture, respectively.

Objectives

The main objectives of aquaculture are:

- To boost the economy of the country by way of increasing per capita fish production and income.
- To generate employment opportunities for the unemployed and under employed persons.
- To utilize full potential of the natural water resources available in the country.
- To uplift the socio-economic status of the farmers.
- To earn foreign exchange revenue by transport of fish to foreign countries.

- To boost the ornamental fish industry for overall economic development.
- To culture larvicidal fishes for control of mosquito larvae.
- To increase production of protein food, in the form of fish, and fulfill the increasing demands of nutritious food.

Levels of Aquaculture

Depending on the intensity of operation and degree of management, aquaculture practices are classified into following four levels:-

- Extensive aquaculture
- Semi-intensive aquaculture
- Intensive aquaculture
- Super intensive aquaculture

Extensive Aquaculture

In extensive level of aquaculture, low stocking densities of 2000-5000 carp fingerlings are used and no supplemental feed is given. Fertilization may be due to stimulate the growth and production of natural food in the water. In such types of culture system, carp culture does not require water exchange during culture period. The ponds used for extensive aquaculture are usually large. The production is generally low, less than 0.5 ton/ha/yr in the case of carps.

Semi-Intensive Level

Semi-intensive aquaculture uses medium size ponds 0.5 ha each with comparatively higher stocking densities than extensive aquaculture (5000-10000 carp fingerlings/ha). Supplementary feeding is done in moderate amounts. In carp culture, water replenishment is done once or twice a month @10 per cent. The production averages around 3-7 tons/ha/yr of carps.

Intensive Level

In intensive level of aquaculture, the pond size is generally small (about 0.2 ha approximately) with very high density of culture organisms *i.e.* 20000 to 25000 carp fingerlings/ha are stocked. The system is totally dependent on the use of formulated feeds. Feeding of the stock is done at regular intervals. Water replacement under intensive culture is effected on a daily basis. Production under intensive level of aquaculture is much higher, for example, about 12 to 15 tons/ha/year in carp culture.

Super-Intensive Level

Super intensive aquaculture needs running water supply and complete daily water exchange is performed. This system is mostly practiced in cement tanks, fiberglass tanks and raceways *etc.* which are fitted with high efficiency biological filters for continues recirculation of water. The size of the tank ranges between 50-100m^3. The cultured organisms are fed with high quality formulated feed. The feed is given through demand feeders. The water quality is regularly monitored with electronic gadgets. Stocking density ranges between 40,000 to 50,000 carp fingerlings/ha. The production ranges between 15-20 tons/ha/yr in case of carps.

Factors to be Kept under Consideration for doing a Successful Fish Culture

- ☆ The major factors to be considered are soil type, topography and water supplies.
- ☆ The soil type influences how well the ponds will hold water; soils which have the water holding capacity, are preferred for pond construction because it prevents leakage. The good quality soil containing a lower limit of 20 per cent clay is necessary for making ponds.
- ☆ The topography determines the size and shape of the ponds. The ponds should not be more than 10 feet depth. For better management, the size of the ponds should not be too much bigger. Sites, where huge individual ponds could be built, can be divided into smaller ponds built in series. The ponds should be rectangular in shape
- ☆ The availability and quality of water determine where and what type of pond should be made. Growing and harvesting are more challenging in case of erratic water supplies. A perennial water source is very much necessary for the successful fish culture. Source of water may be canal, stream, river, lake, or bore well.

Characteristics of a Candidate Species

The species to be selected for aquaculture should have following characteristics:

- ☆ It should have high growth rate
- ☆ It should have capabilities of efficiently utilize and convert the organic production of the water into fish flesh
- ☆ It should be compatible with other species under culture
- ☆ It should be hardy to live in changing physico-chemical conditions such as temperature, pH, turbidity, carbon dioxide and dissolved oxygen
- ☆ Able to reproduce under confined conditions
- ☆ It should be easy to handle and harvest
- ☆ It should have good market demand

In carp culture, usually three Indian major carps *viz. Catla catla, Labeo rohita and Cirrhinus mrigala* and three exotic carps such as *Hypophthalmichthyes molitrix, Ctynopharyngodon idella* and *cyprinus carpio* are selected for culture. This is mainly because of their fast growth and compatibility among each other.

Different Systems

Some Important Cultural System of Aquaculture are listed here as under:

- ☆ Pond culture system

- Raceway culture system
- Cage culture system
- Pen culture system

Aquaculture is practiced through various culture systems/methods. Freshwater aquaculture is carried out in fish ponds, fish pens, fish cages, raceways and on a limited scale in paddy fields. However, fish culture in ponds is the oldest form of aquaculture and at present, it is the only culture system which is mainly adopted by the farmers of our country.

Ponds Culture System

It is the most common method of fish culture. Water is maintained in an enclosed area by artificially constructed ponds where the aquatic animals such the finfish and shellfish are reared. The ponds may be filled with canal water, rain water, bore well water or from other water sources. The pond must be constructed after proper site selection. The climate, topography, water availability and soil quality of the region influence the character of the fish pond. Detailed knowledge regarding different types of fish ponds is a prerequisite for a profitable business in fish culture. A fish farm comprises of different types of ponds namely nursery ponds, rearing ponds, production ponds and breeding ponds *etc.* The number and dimensions of these ponds mainly depends upon the water resource, variety and size of fish to be cultured and type of management. A typical fish pond is a drainable water body with an inlet for the entry of water from water source and an outlet for draining the pond during harvest.

Types of Fish Ponds

Nursery Pond

Nursery ponds are smaller (0.02-0.06 ha) and is mainly prepared to nurse the hatchlings for a period of about two to three weeks *i.e.* until they become fry (2.5-4.0 cm). The depth of the water column may be between 1.0 1nd 1.5 m. The maximum stocking density of hatchlings is about 10 millions/ha. However these ponds are used as nursery only for a short time, they could be used three or four times in a single breeding season. During the other seasons, the nurseries can also be used as production ponds.

Rearing Ponds

Rearing ponds are fairly larger than nursery ponds and sizes usually range between 0.06 and 0.1 ha. In these, the fry are grown for about two to three months or until they attain fingerlings stage (4-10 cm). The depth of water column may be between 1.5 and 2.0 m. Like nursery ponds, when rearing ponds are not in use for rearing purpose, they can serves as production ponds

Production Ponds

In production/stocking ponds, the fingerlings are raised to marketable size fish. The size of this pond varies from 0.1 to 2.0 ha, as ponds larger than 2 ha are not suitable for efficient management. In production ponds for carp culture, the depth of water column should be between 2 and 2.5 m.

Breeding Ponds

These ponds are only needed for breeding purposes. These are used to stock brooders of the fish species to breed. In case the source of water is turbid, a small sedimentation pond or a filtration system may also be constructed to filter the water before its direct entry into the fish ponds. If areas of water scarcity and high seepage are to be utilized for fish farming, cemented ponds may be constructed there. However, such ponds should be treated/overlaid with a soil bed/cover of 30-50 cm soil, in order to give the natural substratum with rich organic matter for higher production and growth.

General Management Practices for Fish Culture

The management practices of fish ponds includes mainly the clearing of algal blooms and weeds, eradication of unwanted animals including predatory and weed fishes by poisons, liming, fertilization, seed stocking, regular supply of required supplementary feed, regular monitoring of water quality parameters and fish livestock health as well as harvesting of table size fishes for marketing.

Clearance of Algal Blooms and Aquatic Weeds

During fish culture, algal blooms may appear. With the climate change, frequency of such activities increases. During night hours, these algal blooms along with different aquatic weeds liberate good amount of carbon dioxide and absorb the dissolved oxygen content available in the fish pond. It is reported that higher carbon dioxide concentration may be responsible for large scale fish mortality. They also compete with fish for nutrients as well as space and also create obstruction in free movement of fishes. Most effective method for the control of algal blooms is introduction of algae feeding fish such as silver carp. Aquatic weeds can be controlled either by manual/mechanical, chemical or biological methods. Mechanical method can be used against emergent and submerged weeds. When aquatic weeds are not controlled by manual/mechanical method, then they can be remove by applying chemicals. Aquatic weeds can also be controlled by biological method *i.e.* by using fish like grass carp @ 150 to 200 kg/ha in culture ponds. These fishes give production without damaging the natural pond ecosystem

Eradication of Predators and Weed Fishes

Presence of undesirable fishes may pose problem as they compete for food, oxygen, space and also prey upon the small seed stocked for culture. Such fishes may be controlled by repeated netting operations, application of mahua oil cake or complete pond drying (if possible).

Control of Predatory Aquatic Insects

Various types of insects either as larvae or adults may not only prey on fish hatchlings but may also compete for food. The most important insects which create problems to fish culture include the water strider, backswimmers, water scorpions, giant water bugs, diving beetles and nymphs of dragonflies. Repeated dragging through water with a fine meshed cloth net may help in the removal of these insects to some extent. However, application of cheap oil soap emulsion (3:1) has been found to be most effective for the control of aquatic insects.

Liming

Liming is done for maintaining soil and water pH. Besides, lime also acts as fertilizers. It should be done one week before fertilization. For a pond having pH around slightly alkaline, dose of lime should be 250-300 kg/ha/year.

Fertilization

We should fertilize the pond for production of natural food organisms. Fertilizers may be organic (cow dung@10-20 ton/ha/year) or inorganic (urea@300-350kg/ha/year, single super phosphate@200-300kg/ha/yr).

Stocking of Fish Seed

Time of stocking and stage of fish species to be stocked is very important in this era of climate change. One should keep in mind about the time, amount and ratio of stocking. Stock the fish early in the morning or evening or during cloudy days, because at that time water temperature is minimum. Packed container should be kept for about half an hour in the pond water to equate the temperature. For production pond, stocking density for semi-intensive level of aquaculture should be 5000-10000 fingerlings/ha. There is always an optimum stocking rate in a particular situation, which gives the highest production and the largest fish. Under crowded condition at a higher stocking rate, fish may compete severely for the food supply, and thus suffer stress due to aggressive interaction. Fishes under stress eat less and grow more slowly. In static pond waters, too much of excretory products of the fish livestock suppress the growth rate of stock. However with efficient removal of such excretory products by the proper water circulation or aeration of the pond water, the fish production per unit area can be increased. Depending on the natural productivity of the pond, stock the different varieties of fish species in proper ratio. In case of three species combination, stocking ratio should be catla (40 per cent), rohu (30 per cent) and mrigal (30 per cent) while in case of six species combination of IMC and exotic carps, stocking ratio should be catla (15 per cent), silver carp (15 per cent), rohu (20 per cent), grass carp (15 per cent), mrigal (15 per cent) and common carp (20 per cent). However depending on the seed availability and other local conditions, the stocking ratios can be changed.

Supplementary Feeding

Feed the fish early in the morning and/or evening time. Artificial feed can be prepared either from animal source (fish meal, shrimp meal, silkworm pupae, poultry waste *etc.*) and/or vegetable source (groundnut oilcake, mustard oil cake, soybean oil cake, rice bran, wheat bran, wheat flour *etc.*). Supplementary feed should be given daily @1-3 per cent of estimated body weight of fish livestock present in the fish pond.

Health Management

Sometimes, due to high stocking densities and polluted feed or water, fish is prone to diseases. Most common symptoms of diseased fish are sleeping sickness, dots on body part, fin rottening, swelling of body, abnormal swimming, loosening of scale, gill rottening and fish comes to water surface. Precaution should be taken that diseased fish is not stocked, use healthy seed for stocking. Periodic sampling of fish can be done from pond for checking/supervise the fish health and growth. Always use aqua medicines with consultation of fisheries experts.

Harvesting

Under favorable conditions, Indian major carp as well as exotic carps attain the marketable size within one year time. After attaining the marketable size fish stock, lower the water level of pond by draining the excess water and harvest the stock with the help of a suitable net.

The Concept of Climate Change

Climate change refers to changes in climate brought about by anthropogenic activities and natural variation that alters the composition of the global atmosphere observed over comparable period of time (IPCC, 2001). It is expected to result to increased frequency and severity of droughts, floods and extreme weather events that adds stress on water resources, food security, health, infrastructure and thus overall development (Kalejaiy Matti *et al.*, 2010).

Causes of Climate Change

Climate can change because of natural processes (bio-geographical) and human activities (anthropogenic). The natural processes are the astronomical, the extraterrestrial factors and volcanic eruption. On the other hand, the anthropogenic factor in climate change involves human activities that either emit large amount of greenhouse gases (GHG); carbon dioxide (CO_2), methane (CH_4), chlorofluorocarbons (CFCs) and nitrous oxide (N_2O) among others into the atmosphere that depletes the ozone layer or activities that reduce the amount of carbons absorbed from the atmosphere. The human factors that emit large amounts of greenhouse gases include industrialization, burning of fossil fuel, gas flaring, urbanization and agriculture. On the other hand, human activities that reduce the amount of carbon sinks are deforestation, alterations in land use, water pollution and agricultural practices.

Impact of Climate Change on Fisheries and Aquaculture

The impacts of the accumulation of GHGs in the atmosphere and water relate to a number of physical phenomena including gradual changes in water temperature, acidification of water bodies, changes in ocean currents and rising sea levels. These physical changes affect ecological functions within aquatic systems and the frequency, intensity and location of extreme weather events. The fish productivity mainly rely on the health and functioning of the ecosystems on which it depends for food, habitat and even seed dispersal. Increased frequency of extreme weather events will also affect the safety of fishers, damage homes, services and infrastructure, particularly in coastal areas (IPCC, 2007) and will also damage many coastal ecosystems including mangroves and reefs. Changes in rainfall will cause a variety of changes in water availability ranging from droughts and shortages to floods and it may also responsible for reduction in water quality. Further, the salinization of groundwater supplies and the movement of saline water upstream in rivers caused by rising sea levels will threaten inland freshwater aquaculture (IPCC, 2007). Increased run-off bringing in nutrients from sewage or agricultural fertilizers may cause algal blooms which in turn lead to reduced levels of dissolved oxygen and 'fish kills'. Rising temperatures similarly reduces level of dissolved oxygen and increases metabolic rates of fish, leading to increases in the cases of sudden fish deaths, declines in production or increases in feed requirements and also increase risk and spread of disease (FAO, 2008). Coastal aquaculture will be exposed to major economic losses from extreme weather events and red tides, the frequency and severity of which are likely to increase. However, not all of the changes will be negative. As sea levels rise, flooding of low lying areas and salinization of groundwater and soil will create ideal conditions for aquaculture in many areas, while simultaneously rendering them unsuitable for regular agriculture.

Climate Smart Aquaculture (CSA)

The aim of Climate Smart Aquaculture (CSA) is to support food security by considering various adaptation and mitigation steps. CSA addresses the challenges of building synergies between the related objectives of climate change mitigation, adaptation and productivity and income increase, Climate smart fisheries and aquaculture will require: improving efficiency in the use of natural resources to produce fish and aquatic foods in a sustainable manner;

Adaptation Strategies

Adaptation is defined as activities that aim "to reduce the vulnerability of human or natural systems to the impacts of climate change and climate-related risks, by maintaining or increasing adaptive capacity and systems resilience" (OECDDAC, 2011)

- ✰ Addressing drivers of vulnerability
- ✰ Diversify sources of household income
- ✰ Promote community based risk management measures to face production failure and price of product
- ✰ Develop innovative risk financing instruments and insurance schemes to reduce climate-related risks
- ✰ Conservation of genetic resources
- ✰ Manage and reduce the disaster risk

Mitigation Strategies

Mitigation promotes efforts to reduce or limit greenhouse gas emissions. The various options to mitigate climate change are:

- ✰ Reducing emission

- Adopting improved aquaculture management
- Avoiding emission
- Reducing post-harvest lost
- Use of fishing practices that adhere to principles of the code of conduct for responsible fisheries
- Replanting mangroves in aquaculture area

Conclusion

Climate change is a serious challenge for all over the world. Impact of climate change has limited sustainable production of fisheries and aquaculture. There is need to adopt climate smart approach because it combines adaptation and mitigation in a way that enhance sustainable fisheries production in view of climatic change. Also, awareness on the climate smart approach is relatively low. Increased awareness on climate smart approaches in fisheries sector have potentials for enhancing food security and sustainable better livelihood for farmers. Suitable adaptation and mitigation measures should be site specific so that they respond to anticipated changes of rainfall and temperature in that particular area.

References

Anyanwu *et al.* (2014). Climate Change: Impacts and Threats on Freshwater Aquaculture. Journal of Fisheries and Aquatic Science. Vol. 9 (5): 419-424.

FAO (2008). Climate change for fisheries and aquaculture. Technical background document from the Expert Consultation, 7 to 9 April 2008, FAO, Rome. Paper presented at 'Climate Change, Energy and Food', High level Conference on Food Security: The challenges of climate change and bioenergy, Rome, Italy (3-5). Available at: ftp: //ftp.fao.org/docrep/fao/meeting/013/ai787e.pdf. Date accessed: 6 December 2010.

FAO Options for decision makers (2008). Workshop on Climate Change and Fisheries and Aquaculture, FAO Headquarters, Rome, FAO, Rome, Italy; 7(9): 5. Available at:

http: //www.fao.org/fileadmin/user_upload/foodclimate/pr esentations/ fish/ OptionsEM7.pdf. Date accessed: 6 December 2010.

Kalejaiy-Matti, R.B., S.A. Nassar, H.O. Audu, R.B. Balogun, R.C. Nwoga and B.G. Garba (2010). Climate change: Causes, implications and corrective strategies. Proceedings of the 44th Annual Conference of the Agricultural Society of Nigeria, October 18-22, 2010, LAUTECH, Oyo State, Nigeria, pp: 1452-1457.

IPCC Climate Change (2007): Synthesis Report Contribution of Working Groups I, II, and III to the Fourth Intergovernmental Panel on Climate Change. Core Writing Team: R.K. Pauchauri and A. Reisinger, eds. IPCC, Geneva, Switzerland, 8.

Onada Olawale Ahmed and Ogunola Oluniyi Solomon(2016). Climate smart aquaculture: a sustainable approach to increasing fish production in the face of climate change in Nigeria. International Journal of Fisheries and Aquatic Studies; 4(3): 444-448.

OECD-DAC (2011). Tracking aid in support of climate change mitigation and adaptation in developing countries, 2011.

Chapter 11

Role of Microbial Consortium in Bio Waste Management

Bhabani Prasad Mondal[1]*, *Rabi N. Sahoo*[1], *Bharpoor S. Sekhon*[2], *Sandeep Sharma*[2], *Sonia Devi*[1] and *Ranu Bhardwaj*[1]

[1]*Division of Agricultural Physics, ICAR-Indian Agricultural Research Institute, New Delhi*

[2]*Department of Soil Science, Punjab Agricultural University, Ludhiana, Punjab*

In this 21st century, every country produces a huge amount of bio wastes or organic wastes due to improvement in production systems. These wastes should be treated properly and quickly for their effective management purposes. In order to enhance the degradation rate of bio wastes, inoculation of microbial consortium consisting of various microbial species to those bio wastes is essential. These microbial species especially bacteria, fungi, actinomycetes are very much effective in eliminating the harmful pathogens and accelerating the decomposition process and thus enhancing the efficacy of bio waste treatments. The composting mediated by microbial communities has gained significant popularity in bio waste management. Cellulolytic microorganisms can expedite the rate of decomposition of organic wastes, especially the lignocellulosic components of organic wastes. Molecular techniques like PCR-DGGE (polymerase chain reaction-denaturing gradient gel electrophoresis), microarray analysis can be employed to find dominant microbial species significantly enhancing the rate of composting. The inoculation of composite microorganisms increases the degree of polycondensation, aromatization and humification of waste products as determined by spectroscopic techniques. The inoculation of thermophilic fungal consortium (*Trichoderma viride*, *Aspergillus niger* and *Aspergillus flavus*) with weekly turning frequency could effectively degrade the organic fractions of municipal solid wastes. The cellulolytic bacterial consortium also plays vital role in promoting faster decomposition of organic wastes through reducing the C: N ratio of the used substrates. Therefore, the consortium of microorganisms can effectively accelerate the rate of degradation of different types of organic wastes and turn them into useful end products. These end products are very much useful not only in suppression of soil borne disease but also in enhancing the crop growth and productivity through improving soil fertility and microbial activity.

Introduction

Improvement in production systems, rapidly growing population and urbanization are mainly responsible for production of huge amount of biological wastes including agricultural wastes, municipal solid wastes, food wastes, kitchen wastes *etc.* which increase the emission of noxious gases to the environment (Leow *et al.,* 2018).The production of biodegradable wastes varies from one country to another country. India nearly generates more than 700 million tons of organic wastes annually from different sources (Pan *et al.,* 2012) whereas, China produces near about 96 million tonnes of kitchen wastes per year (Zhou *et al.,* 2014).

Most commonly used methods of disposing organic wastes in disposal yards lead to soil pollution and occurrence of various diseases to plants, cattle and other dairy animals and also to human beings. Besides burning of organic wastes especially agricultural wastes or agro-residues within the field is considered as a cheap and easy means of disposal of excess residues. But, this practice appends to air pollution, increases soil erosion, noxious gas emission causing respiratory problems and increases the fog incidences in the cities. Direct incorporation of agro-residues like rice straw in field solves the problem of air pollution but it is not feasible due to the short time gap between harvesting of rice and sowing of wheat. Besides, it involves additional cost of labour, irrigation and extra tillage. Along with this, rapid expansion of the cities/towns with massive migration of population from rural to urban centres caused considerable increases in per capita of municipal solid waste (MSW) generation. The increasing rate of solid waste generation, limited landfill spaces and more stringent environmental regulations for new landfill sites and incinerators have increased the waste disposal fees especially in developing countries. Therefore, municipalities and local governments are under heavy pressure to find sustainable and cost-effective solid waste management practices (Saha *et al.,* 2010).

Biological treatment, such as composting of organic wastes mediated by microbial consortia is an eco-friendly and economically viable solution (Wei *et al.,* 2007). Generally, the composting process comprises of multiple steps with various microbial species in each step to generate useful product humus or compost. Highly nutritious compost can be used for enhancing the crop growth and production by improving the soil fertility status. Besides, compost has the potential to suppress the soil borne pathogens through various processes mediated by microbes. It is also essential to characterize the dominant microbial species present during the processes of composting. Molecular techniques particularly PCR based techniques assist in their characterization. Keeping all the points in view, this chapter will describe briefly regarding the various aspects of composting, associated microbial species, their molecular characterization and their efficiency in plant disease suppression and growth promotion.

Composting

Composting is considered as a key process of degrading the bio wastes through utilizing the potential of microbial communities under ambient conditions. Generally, composting is the process of biological transformations of bulky organic wastes to concentrated products with the help of microbial enzymes, which are effective in hydrolysing the complex macro molecules into simpler molecules (Al-Dhabi *et al.*, 2019). This microbiological process narrows down the C:N ratio of the substrate used for composting. Generally, composting consists of three phases: (i) an initial decomposition phase; (ii) a thermophilic phase of intense microbial decomposition and (iii) a maturation phase. Rapidly multiplying thermophilic bacterial species dominate during the thermophilic phase.

Importance of Compost

- Compost improves the soil aggregate stability and water holding capacity.
- It improves the nutrient supplying capacity of soil through increasing the cation exchange capacity (CEC) of soil system.
- It improves the growth of beneficial microorganisms and thus improving soil fertility status.
- It improves the crop growth through enhancing the secretions of growth promoting substances by microbes.
- It helps to suppress the soil borne pathogens and harmful diseases.

Factors Affecting Composting

Various factors affecting the composting processes, which are described below:

Aeration

Composting is a biological oxidation process. Microbes involved in decomposition of organic wastes need oxygen (O_2) for their respiration. During composting, the O_2 content of the composting pit should not below 18 per cent.

Temperature

High temperature is necessary for good composting. Temperature liberated through organic matter (OM) decomposition enhances the process of composting (Finstein and Morris, 1975). Optimum temperature for composting ranges from 45-55°C.

Moisture

Water promotes the clumping of the composted materials through reducing the structural stability of the materials. Optimal moisture content for composting varies from 40-60 per cent.

pH

A high value of pH of the used substrates during initial phase composting increases the rate of decomposition.

C : N Ratio

If the initial C: N ratio of the used materials is greater than 35:1, the rate of composting will be accelerated.

Condition of Starting Materials

The speed of composting (biological oxidation) is directly proportional to the amount of surface exposed to the microbial community. Smaller the particle size of the substrate molecules, the better will be the degradation due to more surface area.

Substrates used for Composting

The following compounds are mainly used for microbial degradation during composting such as lignin, cellulose, hemicellulose, glycogen, pectin, sucrose, alkaloids, chitin *etc.*

Evaluation of Compost Maturity

C : N Ratio

If the C: N ratio is less than or equal to 25:1, it indicates the maturity of the compost.

Variation in Humic and Fulvic Acid Content

During composting, the amount of humic acid increases and fulvic acid decreases (Sugahara and Inoko, 1981).

Germination Index (GI)

A GI index value of more than 80 per cent gives the indication of compost maturity along with the disappearance of the phytotoxicity of the used biological waste materials (Huang *et al.*, 2006).

E4/E6 Ratio

It refers to the ratio of optical densities of humic and fulvic acids at around 460 and 660 nm respectively. It reflects the degree of polycondensation, aromatization and the compost maturation (Said-Pullicino *et al.*, 2007). Besides this, the degree of humification or composting can also be determined through analysing the fluorescence spectra of humic and fulvic acids. Fluorescence excitation-emission spectra (EEM) of humic substances under different treatments during composting have been analysed by Song *et al.* (2014). Their results indicated an increase in aromatization, polycondensation and degree of humification during biological degradation procedure. Xi *et al.* (2012) investigated the Fourier Transformed Infrared (FTIR) spectra of humified substances to characterize various wavebands or spectral regions associated with the maturity of compost. The characteristics features of the spectra are mainly associated with O-H stretching of O-H groups, at 3399 cm^{-1}, aliphatic C-H stretching at 2974 cm^{-1}, and C=N stretching of amides in proteins and peptides at 2375 cm^{-1}, aromatic C=C skeletal vibrations at 1662 cm^{-1}; N-H deformation and C=N stretching of amides, 1546 cm^{-1}. (Droussi *et al.*, 2009). The peak associated with the spectra indicates the extent of aromatization and the completion of composting.

Microorganisms Associated with Biological Degradation

The important microorganisms related to the degradation of organic substrates or biological waste products are tabulated in Table 11.1.

Table 11.1: Microorganisms Involved in different Phases of Composting

Polygenetic Group	*Genus Species*	*Ecological Relevance*	*Succession Stage*
α-protobacteria	*Pseudomonas putida*	Pathogenic	Early
β-protobacteria	*Nitrosomona seuropea*	Nitrifier	Middle
γ-protobacteria	*Methylomonas methanica*	Methanotroph	Late
Endospore forming bacteria	*Bacillus stearothermophilus*	Bacterium in compost	Thermophilic
Actinomycetes	*Streptomyces thermovulgaris*	Pathogenic	Thermophilic
Zygomycetes	*Mucor michei*	Decomposer	Early
Ascomycetes	*Chaetonium thermophilum*	Soil inhabitant decomposer	Early and late thermophilic
Basidiomycetes	*Armillaria mellea*	Cellulytic and lignolytic	Mesophilic

Several researchers have studied the influence of microbial community on faster decomposition of biological waste products. Some works are cited here which shows the potential of microbial consortium on bio wastes degradation. Awasthi *et al.* (2014) studied the influence of fungal consortium (*Trichoderma viride, Aspergillus niger* and *Aspergillus flavus*) with different turning frequencies on the biological degradation of organic fractions of municipal solid wastes (OFMSW) and reported that inoculation of fungal consortium with weekly turning frequency efficiently degraded OFMSW. Song *et al.* (2018) investigated the impact of anti-acidification microbial consortium (AAMC) on carbohydrate metabolism and concluded that the inoculation of AAMC could expedite the carbohydrate degradation process and improve the efficacy of composting. The AAMC was mainly comprised of *Dysgonomonas sp., Pseudomonas caenistrain, Aeribacillus pallidus* strain, *Pseudomonas sp., Lactobacillus salivarius* strain, *Bacillus thuringiens is* strain and *Bacillus cereus* strain. Organic waste decomposition is mediated by various extracellular and intracellular enzymes, produced by various microorganisms, inhabited in soil. In the study of Song *et al.* (2018), they have described several enzyme-based metabolic pathways such as acetic acid metabolic pathway, acetate metabolic pathway, propionic acid metabolic pathway and the enzymes related to those pathways such as pyruvate dehydrogenase, aldehyde dehydrogenase, acetyl-coenzyme A synthetase, malate synthase, acetyl coenzyme A synthetase, and aconitatehydratase. The AAMC significantly improves the production of those above stated key enzymes and thus catalysing the biological decomposition of cellulose, hemicellulose, lignin *etc.* Thermophilic cellulolytic microorganisms accelerate the degradation of biowastes (Bohacz, 2017). *Thermusreesei*is an efficient fungus to degrade the agricultural wastes through production of 100 g of cellulase enzymes per litre of culture medium or broth (Schuster and Schmoll, 2010). Game *et al.* (2016) have developed cellulolytic microbial consortium with *Bacillus* sp, *Aspergillus terreus* and *Streptomyces* sp. for hastening the rate of degradation of organic wastes and concluded that the used consortium enhanced the microbial activity and reduced the period of composting.

Molecular Characterization of Microbial Species

Molecular characterization of microbial species is essential to confirm the dominance of various microbial communities at various phases of composting. Danon *et al.* (2008) used polymerase chain reaction-denaturing gradient gel electrophoresis (PCR-DGGE) technique to identify the dominant bacterial community during composting of biological wastes. 16S rRNA gene sequence can be applied to monitor the changes in microbial communities during decomposition. Partial rRNA gene clone libraries and microarray of oligonucleotide probes are applied for studying the microbial dynamics at molecular level. The database of rRNA gene sequences are submitted to Gene Bank database after matching with the available database of the National Centre for Biotechnology Information (NCBI) under various accession numbers like EU215227 *etc.* Unweighted-pair-group method with arithmetic average (UPGMA) can be employed for computing a complete linkage of dendrogram or phylogenetic tree of microbial species, associated with degradation of biowastes.

Plant Disease Suppression and Plant Growth Improvement

The end product of bio-waste decomposition is compost, which has the potential to suppress the growth of soil-borne plant pathogens and thereby reducing the application of pesticides to the soil environment. Hyper-parasitism, antibiotic production by microbes, activation of disease resistant genes within the host plants *etc.* are some popular mechanisms mediated by microbes to subdue the harmful soil borne pathogens. Actinomycetes like *Streptomyces* sp. have the capability to mollify the effects of disease causing organisms especially the phytopathogens by production of antibiotics and other bio active secondary metabolites (Al-Dhabi *et al.*, 2019). Phytopathogen *Rhizoctonia solani* can be effectively controlled by the antifungal secondary metabolites secreted by *Streptomyces* sp. (Al-Dhabi *et al.*, 2019). Along with the disease suppression, the *Streptomyces* sp. can biosynthesize various growth promoting substances like indole acetic acid (IAA), siderophore for improving the plant growth. These compounds also enhance root and shoot growth and thus enhancing the nutrient foraging capacity of the crop plants.

Conclusion

Undoubtedly, the use of microbial consortium rather than single inoculum is a viable option to enhance the rate of decomposition of bio wastes. The conventional techniques like chemical and thermal techniques are quite expensive and hazardous to environment also. Thus, from the practical point of view, it can be stated that the microbiological technique is an effective alternative technique to those conventional techniques for bio waste management in terms of economy and energy consumption. Molecular techniques like PCR-DGGE, rRNA analysis are very much crucial for identifying the dominant microbial species at different phases of composting.

Spectroscopic techniques like FTIR spectra, fluorescence spectra can be successfully applied to determine the maturity of compost and the efficacy of composting, inoculated with microbial consortia. Thermophilic fungal consortia and cellulolytic bacterial consortia are very much effective in accelerating the rate of decomposition of various biowastes. The microbial consortium is also efficient in degrading various types of organic wastes, enriching the end product with various types of growth promoting substances like IAA, other phytohormones *etc.* and also suppressing the phytopathogens through secreting antagonistic secondary metabolites.Further studies are also essential to develop effective bio-inoculants for various agro-climatic zones for quality compost production to fulfil the demands of the local farmers.

REFERENCES

Al-Dhabi, N.A., Esmail, G.A., Mohammed Ghilan, A.K. and Valan Arasu, M.,(2019). Composting of Vegetable Waste Using Microbial Consortium and Biocontrol Efficacy of Streptomyces Sp. Al-Dhabi 30 Isolated from the Saudi Arabian Environment for Sustainable Agriculture. *Sustainability* **11(23):** 6845.

Awasthi, M.K., Pandey, A.K., Khan, J., Bundela, P.S., Wong, J.W. and Selvam, A. (2014). Evaluation of thermophilic fungal consortium for organic municipal solid waste composting. *Bioresource technology***168:** 214-221.

Bohacz, J. (2017). Lignocellulose-degrading enzymes, free-radical transformations during composting of lignocellulosic waste and biothermal phases in small-scale reactors. *Science of The Total Environment* **580:** 744-754.

Danon, M., Ingrid H., Franke W, Heribert, I., Yona, C. and Yitzhak, H. (2008). Molecularanalysis of bacterial community succession during prolonged compost curing. Federation of European Microbiological Societies. *FEMS Microbiology and Ecology***65:** 133–144.

Droussi, Z., D'Orazio, V., Hafidi, M., Ouatmane, A. (2009). Elemental and spectroscopic characterization of humic-acid-like compounds during composting of olive mill by-products. *Journal of Hazardous Materials***163:** 1289–1297.

Finstein, M.S. and Morris, M.L. (1975). Microbiology of municipal solid waste composting, *Advances in Applied Microbiology***19:** 113-151.

Game, B., Deokar, C. and Navale, A. (2016). Composting of organic wastes using newly developed cellulolytic microbial consortium. *International Journal of Agriculture, Environment and Biotechnology***9(4):** 525-534.

Huang, G.F., Wu, Q.T., Wong, J.W.C., Nagar, B.B. (2006). Transformation of organic matter during co-composting of pig manure with sawdust.*Bioresource Technology***97:** 1834–1842.

Leow, C.W., Van Fan, Y., Chua, L.S., Muhamad, I.I., Klemes, J.J. and Lee, C.T. (2018). A review on application of microorganisms for organic waste management. *Chemical Engineering Transactions***63:** 85-90.

Pan, I., Dam, B. and Sen, S.K. (2012). Composting of common organic wastes using microbial inoculants. *Biotechnology***2(2):** 127-134.

Saha, J.K., Panwar, N., Singh, M.V. (2010). An assessment of municipal solid waste compost quality produced in different cities of India in the perspective of developing quality control indices. *Waste Management***30:** 192–201.

Said-Pullicino, D., Erriquens, F.G., Gigliotti, G. (2007). Changes in the chemical characteristics of water-extractable organic matter during composting and their influence on compost stability and maturity. *Bioresource Technology***98:** 1822–1831.

Schuster A., and Schmoll M. (2010). Biology and biotechnology of Trichoderma. *Applied Microbiology and Biotechnology***87(3):** 787-799.

Song, C., Li, M., Wei, Z., Jia, X., Xi, B., Liu, D., Zhu, C. and Pan, H. (2014). Effect of inoculation with multiple composite microorganisms on characteristics of humic fractions and bacterial community structure during biogas residue and livestock manure co composting. *Journal of Chemical Technology and Biotechnology* **91(1):** 155-164.

Song, C., Li, M., Qi, H., Zhang, Y., Liu, D., Xia, X., Pan, H. and Xi, B. (2018). Impact of anti-acidification microbial consortium on carbohydrate metabolism of key microbes during food waste composting. *Bioresource technology***259:** 1-9.

Sugahara, K. and Inoko, A. (1981).Composition analysis of humus and characterization of humic acid obtained from city refuse compost.*Soil Science and Plant Nutrition* **27(2):** 213-224.

Wei, Z., Xi, B., Zhao, Y., Wang, S., Liu, H. and Jiang, Y. (2007). Effect of inoculating microbes in municipal solid waste composting on characteristics of humic acid. *Chemosphere***68:** 368–374.

Zhou, H., Meng, A., Long, Y., Li, Q., Zhang, Y. (2014). An overview of characteristics of municipal solid waste fuel in China: Physical, chemical composition and heating value. *Renewable and Sustainable Energy Reviews* **36:** 107–122.

Chapter 12

Impacts of Stubble Burning and its Management

Vinita Parte[1], Vishnu K Solanki[1], Anjana Kujur[2] and J.S. Ranawat[3]*

[1]JNKVV, Jabalpur, Madhya Pradesh
[2]RVSKVV, Gwalior, Madhya Pradesh
[3]Chaudhary Charan Singh Hisar Agricultural University, Hisar, Haryana

Stubble is defined as the leaves, straws and roots of cereals left in the field after the harvest. The stubbles constitute about 50-75 per cent of the total cereal biomass produced in a season. Rapid and cost-free destruction, reduction of some weeds, pests and diseases, facilitation of soil tillage, time-saving in growing second crops and preparation of seedbed at low cost encourage farmers for stubble burning. However, stubble burning results in death of various living creatures, loss of organic matter in the soil, deterioration of soil structure, destruction of useful microorganisms, air pollution, fires, soil erosion and some health problems (asthma, bronchitis *etc.*). Instead of burning the cereal stubble which is an important feed source for livestock in barns, farmers should be encouraged to use stubbles as animal bedding, organic matter through incorporating them into the soils, as raw materials in pulp and paper industry and as a source of fuel. (Ozge Doganay ERBAS KOSE *et al.*, 2018). Post-harvest plant biomass, in other words, stubble is commonly burnt by the farmers. Such a case is defined as stubble burning and it has several negative impacts on soil quality and fertility. The following factors force the farmers to burn stubbles in cases where stubble and stalks are not wanted and economically used. (Kara and Sezer, 1992; Silme *et al.*, 2015).

- It is quite easy, free of cost and fast elimination process.
- Reduction in self-growing cereals and weeds.

- ☆ Reduction in snail, centipede, ticks, wireworms, spiders and other insect populations.
- ☆ Reduction in some diseases.
- ☆ Convenience in soil tillage, energy saving because of reduced processes.
- ☆ Time saving for second crop, minimization of drilling problems in no-tillage systems developed for economic productions.
- ☆ Stubbles affect the performance of seeding machines through plugging burying legs, preventing.
- ☆ Penetration into the soil and tangling around the other parts of the machines.
- ☆ Seedbed preparation at a lower cost.
- ☆ Facilitate field preparation for the subsequent seeding period since degradation of plant residues is quite slow and time-consuming in soils of humid and cold regions.
- ☆ High yield expectations.

It is well known fact that Clouds of ash and smoke can easily travel for more than thousand kilometers and it can create an obstinate clouds. Also from the smoke, smog is formed which can increase the levels of pollutants in the air, which in turn cause breathing problem and other health issue like irritation in eyes *etc.* Once the pollutant is released in the atmosphere, it changes the chemical composition of the air quality affecting the human health and recurrent burning of husk may cause to the formation of the brown clouds which affects the local air quality, atmospheric visibility and global warming too. Most of the farmers believed that stubble burning is a quick, easy and cheap method as by these methods all the unwanted husk will gets destroyed along with other unwanted grasses. Also many farmers believed that fire actually return the nutrients of the land which is a misconception or rather we can say lack of knowledge due to awareness and illiteracy. In practical, Burning of husk on ground actually destroys the nutrients value of the soil due to which soil become less fertile because the heat generated from the stubble burning penetrates into the soil, causing loss of the moisture, a useful microbe which in results affects the quality of the soil. Apart from this natural nutrients and bacteria which help in the rejuvenating of the soil were also destroyed. Not only this stubble burning also the reason for the loss of precious nutrients.

Fire Effects on Soil Biological Activity

So far, very little information on the effects of fire on soil biological life is achieved. It has been reported that the fire gradually decreased soil organic matter and biological activity. One reason suggested for reducing the activity of micro-organisms, destroying them by fire. Add organic matter to the soil of the cultivation of plants and micro-organisms in the soil underneath again and areas of soils that have not been exposed to fire. Provided a food source for soil microbe's fire soil organic matter levels are severely reduced. Studies show that about 80 per cent of plant residues by microbial oxidation process and the remaining 20 per cent is converted into organic compounds joins. The cause of the microbial biomass production, or compounds that are easily degraded by microorganisms and increase soil biological activity.

When remains are cremated remains about 60 per cent of the amount of its immediately (carbon monoxide) becomes and the remaining 40 per cent of this amount, 32 per cent will be added to the soil microbial processes and their breathing to become the 8 per cent of the organic matter is converted. Carbon in the ash cannot exist as a food source for the microorganisms and act as a support for them. Thus, the activity of micro- organisms and population declines and soil biological processes are disrupted. During a test plant debris and ashes from the burnt remains were added to pots containing loamy soil in the laboratory and field capacity for 10 days at 24 ° C and temperature were maintained at the end of the microorganisms was evaluated in pots. Add plant remains much more than adding ash was due to microbial activity. It must be said that the plant remains in the treatment of nitrogen were added to a large amount of nitrogen in the soil due to very high carbon to nitrogen ratio (C/N) was immobile, but the ash was added in the mixture of soil available nitrogen was more due to the lack of soil microbes use of ash as a source of food. Framed population declined as a result of micro-organisms and nitrogen rather than the plant was established by soil microbes. (Einallah Hesammi *et al.*, 2014)

Fire Effects on Crops

Almost all sources have reported that the next crop will increase the burning remains. Burning residue that remains of nutrients and need to be released directly in the biological cycle of the plant and rotting food shop and a significant increase in the next crop to be viewed. It should be noted that this term is due to the detrimental effects that this has on soil properties described in the previous discussion in time with the decrease of product due to the accumulation will face minerals, because of the product a temporary increase In the past, farmers have shifted to operate without being aware of the devastating effects of fire, but have little today to clarify the damaging effects of this action is by arable better successor. Effects of organic inputs over time on soil physical properties (soil aggregate stability, soil bulk density, water retention *etc.*) was reported by several authors (Zhang and Peng, 2006; Fuentes *et al.*, 2009). The organic products additions to the soil increase aggregate stability by a factor of 1.1-10.0, and this increase is related to the decomposition dynamics of the inputs (Abiven *et al.*, 2009).

The Effect of Fire on Soil Properties

Structure, aggregation, mechanical strength and bulk density: Fire destroyed remnants of organic matter, and this directly affects the structure and soil granulation side effects. Due to the correlation between soils organic matter disappears and low soil pore spaces. Favorable soil aggregation fires (due to reduced organic material) and disappears over time traveling machine of soil is pressed. In addition, the decrease in organic matter, soil bulk density and soil porosity increase and decrease the side effects on the growth of plants and micro-organisms, because in these Terms of air conditioning and decreases the soil Gas Exchange. It also reduces the adhesion between soil particles causes the formation of large lumps during plowing tillage that they should do more to grind that is the subject of increased soil compaction and tillage.

Major Pollutants Emitted during Crop Residue Burning and its Source (Singh *et al.*, 2008)

SPM (Suspended Particulate Matter) (PM100)

Incomplete combustion of in organic material, particle on burnt Soil

FPM (Fine Particulate Matter)

Condensation after combustion of gases and incomplete combustion of organic matter

CO-Incomplete combustion of organic matter

NO_2, N_2O - Oxidation of N_2 in air at high temperature

O_3 - Secondary pollutant, form due to Nitrogen Oxide and Hydrocarbon

CH_4/Benzen - Incomplete combustion of organic matter

Different Technologies to Manage Paddy Crop Residues and Reducing Stubble Burning

Due to ever growing predicaments linked with crop residue burning in the north western states of India, several proposals and techniques have been developed for its appropriate handling in the past years by different agencies were as follow:

Collection of Residues for off Farm Uses

In some parts of India straw reaper are in practice to collect the straw from the field and is only popular in wheat straw collection instead of rice because of its economical use for feeding animals. For removal and collection of straw after combine harvesting and using the residues for off farm works; straw baler machines is very promising technology and commercially available. These balers, however, recover only 25–30 per cent of potential straw yield after combining, depending upon height of plant cut by combines. Baler makes rectangular or round bales by collecting the loose straw from the ground. Thus baler provides a solution for residue management in an environmental friendly way.

Crop Residue Retention on Surface Soil

Retention of crop residue on surface soil as mulch increases the least soil temperature in winter through reducing upward heat flux from soil and declines soil temperature during summer due to shading effect. Pathak *et al.*, stated that crop residues play a significant role in betterment of soil acidity by releasing bases during the decomposition of crop residues with higher C: N, and improves soil alkalinity through application of residues from lower C:N crops such as legumes, oil seeds and pulses. Other benefits of retention of crop residues on soil surface includes lesser weed growth which saves weedicide cost, improves physico-chemical and biological properties of soils and recycling of plant nutrients for lowering fertilizer use in the next successive crops. Mulching with crop residue also increases infiltration, reduces formation of soil crust and runoff. Resource conservation technologies (RCTs) based farm machinery provides a better promise in managing paddy residues for improving soil health, productivity, reducing pollution and achieving sustainable agriculture (Palma *et al.*, 2014).

In situ Crop Residue Incorporation

In situ residue incorporation has several positive impacts on soil health such as pH, organic carbon, infiltration rate and water holding capacity (Kumar *et al.*, 2015). It increases hydraulic conductivity, cation exchange capacity (CEC), and reduces bulk density of soil by modifying soil structure and aggregate stability, surface crust formation, water evaporation from the top few inches of soil and prevents leaching of nutrients. It also increases the microbial biomass and enhances activities of enzymes such as dehydrogenase and alkaline phosphatases. Sidhu *et al.* reveal that soil incorporated with crop residues held 5–10 times more aerobic bacteria and 1.5–11 times more fungi than soil from which residues were either burnt or removed.

Another study (Singh, 1992) reveals that, incorporation of paddy straw in soil immobilized native as well as added fertilizer N and about half of the immobilized N was mineralized after 90 days of straw incorporation. Straw and N application alone or in combination increased biomass carbon, phosphates and respiratory activities of the soil. Microbial biomass carbon and phosphate activities were observed maximum at 30 days of straw decomposition. In field trials, incorporation of paddy straw 3 weeks before sowing of wheat significantly increased the wheat yield at Sonepat district in a clay loam soil while no such beneficial effect was observed in a sandy loam soil at Hissar (Singh, 1992).

Use of Rice Residue as Bedding Material for Cattle (Parmod kumar *et al.*, 2015)

The farmers of the state have been advised to use paddy straw as bedding material for cross bred cows during winters as per results of a study conducted by the Department of Livestock Production and Management, College of Veterinary Sciences, Punjab Agricultural University. It has been found that the use of paddy straw bedding during winter helped in improving the quality and quantity of milk as it contributed to animals' comfort, udder health and leg health. Paddy straw bedding helped the animals keep themselves warm and maintain reasonable rates of heat loss from the body. It also provides clean, hygienic, dry, comfortable and non-slippery environment, which prevents the chances of injury and lameness. Healthy legs and hooves ensure enhancement of milk production and reproductive efficiency of animals. The paddy straw used for bedding could be subsequently used in biogas plants.

Uses of Crop Stubble

The crop stubble produced during the harvesting of rice and wheat crops can be used for various alternative uses if it is not burnt. These include use of crop stubble as fodder for animals, use of crop stubble for the generation of electricity, use as input in the paper/pulp industry *etc.* The use of rice residue as fodder for animals is relatively low in Punjab as compared to the wheat stubble. This is because the rice residue is high in silica content which in turn is not good for animal health. However, very often the crop stubble is treated with urea before it is fed to the animals. As per Badve (1991), treating crop residues with 4 per cent urea and 45–50 per cent moisture improves the nutritive value by increasing digestibility, palatability and crude protein content.

Alternative Uses of Crop Stubble

Agricultural waste includes paddy and wheat straw, cotton sticks, bagasse and animal waste. Keeping in view the increasing problems associated with crop stubble burning several initiatives for its proper management have been taken up. Various departments and institutions are promoting alternative uses of straw instead of burning. These include:

- ☆ Use of Rice Residue as Fodder for Animals.
- ☆ Use of Crop Residue in Bio Thermal Power Plants.
- ☆ Use of Rice Residue as Bedding Material for Cattle.
- ☆ Use of Crop Residue for Mushroom Cultivation.
- ☆ Use of Rice Residue for Making Bio Gas.
- ☆ Production of Bio-oil from Straw and Other Agricultural Wastes.

REFERENCES

Abiven S, Menasseri S, Chenu C (2009). The effects of organic inputs over time on soil aggregate stability-A literature analysis. *Soil Biology and Biochemistry*: 41: 1-12.

Badve, V. C. (1991). Feeding systems and problems in the Indo-Ganges plain: Case study. In A.Speedy and R. Sansoucy (Eds.), Feeding dairy cows in the tropics. *Proceedings of the FAO Expert Consultation Held in Bangkok, Thailand*, July 7–11, 1989.

Einallah Hesammi *et al.* (2014). A review on the burning of crop residue on the soil properties.*WALIA journal*, 30(1): 192-194.

Fuentes M, Govaerts B, De Leónc F, Hidalgo C, Dendooven L, Sayre KD, *etc.*hevers J (2009). Fourteen years of applying zero and conventional tillage, crop rotation and residue management systems and its effect on physical and chemical soil quality. *European Journal of Agronomy* 30: 228-237.

Kara, E., Sezer, I. 1992. Aniz Yakma. Ekoloji, 5: 18-22.

Kumar, P.; Kumar, S.; Joshi, L (2015). The extend and management of crop residue stubbles. In Socioeconomic and Environmental Implications of Agricultural Residue Burning: *A Case Study of Punjab, India; Kumar, P., Kumar, S., Joshi, L., Eds.; Springer Briefs in Environmental Science*: Berlin, Germany, 2015; p. 144,ISBN 978-81-322-2014-5.

Ozge Doganay ERBAS KOSE *et al.* (2018). Environmental Impacts of Stubble Burning and Potential Use of Stubbles in Different Industries. *International Eurasian Conference on Science, Engineering and Technology,* November 22-23.

Palma, C.; Humberto, B.C.; De Clerckc, F.; Gaterea, L. and Grace, P. (2014) Conservation agriculture and ecosystem services: an overview. *Agr Ecosyst Environ*.187: 87–105.

Parmod kumar *et al.*, 2015. Alternative Uses of Crop Stubble-chapter; P.77.

Pathak, H.; Saharawat, Y.S.; Gathala, M. and Ladha J.K. (2011). Impact of resource-conserving technologies on productivity and greenhouse gas emissions in the rice-wheat system. *Greenhouse Gases, Sc Tech*.1(3): 261–277.

Sidhu, B.S.; Beri, V. and Gosal, S.K. (1995). Soil microbial health as affected by crop residue management. In Proceedings of National Symposium on Developments in Soil Science, Ludhiana, India pp. 45–46.New Delhi, India; *Indian Society of Soil Science*; Nov. 2–5.

Singh, R. P., Dhaliwal, H. S., Sidhu, H. S., Manpreet-Singh, Y. S., and Blackwell, J. (2008). Economic assessment of the Happy Seeder for rice-wheat systems in Punjab, India. *Conference Paper, AARES 52nd Annual conference, Canberra. Australia*: ACT.

Singh, S., Batra Renu, Mishra, M. M., Kapoor, K. K., and Goyal Sneh (1992). Decomposition of paddy straw in soil and the effect of straw incorporation in the field on the yield of wheat. *Journal of Plant Nutrition and Soil Sciences,* 155(4), 307–311.

Slime, R.S., Gumrukcu, E. ozkan, C.F., Baysal, O. 2015. Aniz Yakman Toprakta Mikroflora Dinamigine olan Etkilen. Elektronic Mikrobiyoloji Dergisi, 13(1): 47: 51.

Zhang B, Peng X-H (2006). Organic matter Enrichment and Aggregate Stabilization in a Severely Degraded Ultisol after Reforestation. Pedosphere 16(6): 699-706.

Chapter 13

Human Health Risks and Soil Pollution

Indu Singh and Bhopal Singh

Assistant Professor,

Department of chemistry, Meerut College Meerut, Uttar Pradesh

Soil pollution occurs when the presence of toxic chemicals, pollutants or contaminates in the soil is in high enough concentrations to be of risk to plants, wildlife, humans add of course, the soil itself. Both organic and inorganic contaminants are presents in the soil. Example of organic contaminants is fuel hydrocarbons, polynuclear aromatic hydrocarbons, polychlorinated biphenyl, chlorinated aromatic compounds, detergents and pesticides. Inorganic species include nitrates, phosphates, and heavy metals. Soil pollution is heavy metals which adversely affect soil quality and human health. Human activity introduces heavy metal such as arsenic, mercury and cadmium to soil through smelting, industry, burning fossil fuels and agricultures *etc.* Soil and human health studies include investigation into nutrient supply through the food chain and routes of exposure to chemical and pathogens.

Introduction

Soil pollution occurs due to the presence of toxic compounds, chemicals, radioactive materials, salts *etc.* in the soil that have adverse effects on plants and animals. Soil is covers by organic and inorganic materials. Mainly organic materials comes from decayed remains of plants and animals and inorganic portion derived from rock fragments which formed over thousands of years by physical and chemical weathering of bedrock. From different ways, soil can become polluted like seepage from landfill, discharge of industrial waste into the soil, percolation of contaminated water into the soil, solid waste seepage, excess of pesticides, herbicides and fertilizers *etc.* Some contaminants are presents in plants which come from pesticides and fertilizers, enter into the food chain and impart adverse effect on human as well as other organisms.

Causes of Soil Pollution

Industrial Activity

Industrial activity plays an important role contributes to soil pollution by manufacturing of different types of products. Some industries are based on extracting various minerals both metallic and non metallic minerals from the earth. By products formed in manufacturing process with main products are contaminated and they are not disposed in proper manner. These industrial wastes are stay in the soil surface for long time and make it polluted.

Agricultural Activities

Some chemicals like pesticides, fungicides, fertilizers, insecticides *etc.* are used in the agriculture. These chemicals are produced in the nature that cannot be broken down in the nature and seep into the soil with water then slowly reduces the fertility of the soil.

Waste Disposal

Domestic wastes comes from home as well as garbage and rubbish wastes like glasses, waste paper, metal cans and fibres *etc.* disposed in to ground and these wastes contribute to soil pollution.

Acid Rain

Various gases are present in the atmosphere in form of primary pollutants and secondary pollutant such as carbon dioxide, sulphur dioxide, sulphur trioxide, nitrogen dioxide, nitrogen trioxide. Chlorofloroethanes *etc.* When rain comes, it mixed with these pollutants and formed acid rain which mixed in the soil. The polluted water dissolve away nutrients present in soil and produced soil pollution because it change the structure of soil.

Effects of Soil Pollution

Sources	*Adverse Effects*
Toxic heavy metals in soil	Irreversible development damage in children, neurological problem causes by lead and kidney and liver damage by mercury
Exposure of Benzene, Polychlorinated biphenyl	Leukemia in children and adults, liver cancer
Herbicides	Kill weeds and harm people including cancer and disorder
Pesticides and insecticides	Wiped out of birds and eagles, hyper activity disorder in humans
Garbage dumps and landfills	Introduce toxins like arsenic into ground water causes illness and birth defect
Nuclear waste dumps	Exposures to harmful radiation causes cancer and birth defects

Sources	*Adverse Effects*
Urban development	Causes mental health issues
Toxic chemicals	Decrease soil fertility which lacks nutrients and causes serious health problem
Polluted soil in drinking water	Cholera and Dysentery, Malaria, Skin and stomach infections

Prevention of Soil Pollution

Small Use of Chemical Fertilizers

Chemical fertilizers are used for enhance the fertility of the soil but in excess amount of these chemical fertilizers could pollute the soil. Chemicals could also destroy the good microorganisms present in the soil. When chemicals runoffs from the soils also cause water pollution

Afforestation and Reforestation

The main causes of soil pollution are soil erosion which is caused by deforestation. Human being needs more space to expand civilization due to over-growing population. We should be promoted afforestation in the barren lands. By afforestation and reforestation, the roots of the plants bind the soil particles and capture good microorganisms in the soil as well as maintenance of the underground water table.

Use of Natural Manure

Natural manure is the best method for maintenance of nutrients in the soil because it is harmless and completely organic. It adds important nutrients to the soil and restores the health of the soil.

Recycle and Reuse Products

Recycle and reuse can reduce waste generation as well as reduce soil pollution. If plastics and other wastes are buried in landfills, these wastes decompose slowly and release toxic substances into the soil that harmful to the health of the soil. We should be parchage glass and other reusable containers instead of throwing away plastic.

REFERENCES

Bech, J. Soil contamination and human health: Part 1-preface. Environmetal Geochemistry and Health. 2020; 42: 1-6.

Brevik, E.C.; Slaughter, L.; Singh, B.R.; Steffan, J.J.; Collier, D.; Barnhart,P. and Pereira, P. Soil and human health: current status and future needs. Air, Soil and Water Research. 2020; 13: 1-23.

Chibuike, G.U. and Obiora, S.C. Heavy metal polluted soil: Effect on plants and bioremediation methods. Applied and Environmental Soil Science. 2014; doi.org/10.1155/2014/752708.

Fernandes, C.L.F.; Volcao, L.M.; Ramires, P.F.; De Moura, R.R. and Junior, M.R.D.S. Distribution of pesticides in agriculture and urban soils of Brazil: a critical review. Environmental Science: Processes and Impacts. 2020; 22: 256-270.

Ferreiro, J.P.; Gasco, G. and Reichman, S.M. Soil pollution and remediation. International Journal of Environmental Research and Public Health. 2018; 15(8): doi: 10.3390/ijerph15081657.

Hou, D.; OConnor, D.; Igalavithana, A.D.; Alessi, D.S.; Luo, J.; Tsag, D.C.W.; Sparks, D.L.; Yamauchi, Y.; Rinklebe, J. and Ok, Y.S. Metal contaimination and bioremediation of agricultural soils for food safety and sustainability. Nature Reviews Earth and Enviroment. 2020; 1: 366-381.

Huang, B.; Yuan, Z.; Li, D.; Zheng, M.; Nie, X. and Liao, Y. Effects of soil particle size on the adsorption, distribution, and migration behaviors of heavy metal(Ioid) s in soil: a review. Environmental Science: Processes and Impacts. 2020; 22: 1596-1615.

Kumar, V.; Pandita, S.; Sharma, A.; Bakshi, P.; Sharma, P.; Karaouzas, I.; Bhardwaj, R.; Thukral, A.K. and Cerda, A. Ecological and human health risks appraisal of metal(Ioid)s in agricultural soils: a review. Geology, Ecology and Landscapes. 2019; Doi: org/10.1080/24749508.2019.1701310.

Labe, T.E. and Agera, S.I.N. Role of forestry in mitigating global soil pollution from toxic heavy metals-a review. Journal of Research in Forestry, Wildlife and Environment. 2017; 9(2): 92-101.

Li, X.N.; Jiao, W.T.; Xiao, R.B.; Chen, W.P. and Chano, A.C. Soil pollution and site remediation policies in china: A review. Canadian Science Publishing. 2015; doi: org/10.1139/er-2014-0073.

Pereira, V.J. and Martinho, D. exploring the topics of soil pollution and agricultural economics highlighting good practices. Agriculture. 2020; 10: doi: 10.3390/agriculture10010024.

Ranieri, E.; Bombardelli, F.; Gikas, P. and Chiaia, B. Soil pollution prevention and remediation. Applied and Environmental Soil Science. 2016; Doi.org/10.1155/2016/9415175.

Rashtian, J.; Chavkin, D.E. and Merhi, Z. Water ad soil pollution as determiant of water and food quality/contamination and its impact on female fertility. Reproductive Biology and Endocrinology. 2019; 17(5): doi.org/10.1186/s12958-018-0448-5.

Rodrigues, S.M. and Romkens, P.F.A.M. Human health risks and soil pollution. Wageningen University ad Research. 2018; 9: 217-250.

Sagar, N. and Narayan, R. Soil Mycobiota influenced by different concentration of basic fuschin dye. IJCRR. 2019; 11(17): 12-17.

Seiyaboh, E.I. and Izah, S.C. Impacts of soil pollution on air quality under Nigerian setting. Journal of Soil and Water Science. 2019; 3(1): Doi: 10.36959/624/430.

Shifaw, E. Review of heavy metals pollution in China in agricultural and urban soils. Journal of Health and Pollution. 2018; 8(18): doi.org/10.5696/2156-9614-8.18.180607.

Steffan, J.J.; Brevik, E.C.; Burgess, L.C. and Cerda, A. The effect of soil on human health: a overview. European Journal of Soil Science. 2018; 69(1): 159-171.

Taddese, S. Municipal waste disposal on soil quality. A review. Acta Scientific Agriculture. 2019; 3(12): 09-15.

Tetteh, R.N. Chemical soil degradation as a result of contamination: A review. Journal of Soil Science and Environmental Management. 2015; 6(11): 301-308.

Chapter 14

Red Mud in Wastewater Treatment Technology

B.M. Porkavi, P. Kalaiselvi, S. Paul Sebastian, V. Davamani, E. Parameswari and R. Jeyarani Haripriya*

Department of Environmental Sciences, Tamil Nadu Agricultural University, Coimbatore – 641 003, Tamil Nadu

Red mud (RM) is an industrial waste produced in large quantities from bauxite during the extraction of alumina. Due to high alkalinity, its disposal produces significant environmental pollution. Red mud can be modified into useful products, such as adsorbents in wastewater treatment. Red mud can be developed as an affordable adsorbent due to its high concentrations of aluminum oxides, iron oxides, titanium oxides, silica oxides and hydroxides, for the removal of various ions from aqueous solution and soils (*e.g.*, metal and non-metal ions, phenolic compounds and dyes) and for the purification of waste gases (sulfides and carbide). This review outlined the utilization of RM in wastewater treatment.

Keywords: Red mud, Wastewater treatment, Environmental remediation, Waste gas.

Introduction

Solid waste materials (by-products) produced from multiple industrial activities are well known to raise one of the most perplexing problems in society. The lack of proper disposal of solid waste, including industrial waste, remains one of the major problems to be solved in many developing countries. To turn the solid waste into 'low-cost adsorbents' for the treatment of wastewater generated from diverse industries is a significant and effective use of solid waste. If solid wastes can be used as low-cost adsorbents, they can have a two-fold environmental pollution control benefit. Firstly, the amount of waste materials could be reduced in part and, secondly, the low-cost adsorbent if produced could minimize the emission of wastewater at a fair cost. Various industrial wastes have been identified as adsorbents for the removal of various forms of toxins from wastewater, such as fly ash, slag, red mud and various types of sludge.

Among various industrial by products, red mud is a substantial solid industrial waste produced in the production of alumina and digestion of bauxite. To produce 1 ton of alumina, approximately 2-3 tons of bauxite is excavated. It was expected that the global stock of RM will cross approximately 4 billion tons in 2019.Red mud has been used successfully by a number of workers as a stabilizing agent for *in situ* fixation of heavy metals in contaminated soils, mine tailings and other wastes. Red mud has found some applications in making different construction materials and ceramic products. Furthermore, metal recovery from red mud is also being developed. The aim of the present review is exploring the adsorption potential of red mud as a low cost adsorbent for the removal of different pollutants from wastewater

Application of Red Mud in Wastewater Treatment

Coagulant

Commonly used coagulants are $Fe^{3+}$and Al^{3+}compounds. Due to high contents of Feand Al in red mud, this is believed to be a promising material for coagulant production (Orešcanin *et al.*, 2001). The coagulant produced from the red mud has a few advantages compared with the commercial Fe and Al salts (Orešèanin *et al.*, 2017). They are,

- ☆ The coagulation/flocculation process could be done in one step without coagulant aid or any other chemical addition.
- ☆ pH measurement during the purification could be avoided due to the negligible pH variations of the treated solution.
- ☆ Allow one step removal of the cationic and anionic species from the wastewater.
- ☆ One dose of coagulant is sufficient for five repeated treatment cycles.
- ☆ Negligible leaching of heavy metals from the waste mud under normally environmental conditions.
- ☆ Safe for transport, storage and handling.

The chemical treatment of red mud will produce coagulants for wastewater treatment due to the presence of high concentrations of calcium, aluminium, iron, and silicon. The treatment of red mud with hydrochloric pickle liquor ($AlCl_3$ and HCl) of bauxite can transform it into a novel composite inorganic coagulant that has shown high efficiency to remove phosphate (Zhou and Haynes, 2011) and better coagulation performance than untreated red mud and commercial polyaluminum chloride coagulant. The composite red mud coagulant is an efficient coagulant in tertiary treatment of wastewaters and to remove nutrients from wastewater and water.

Adsorbent

Among the techniques used in wastewater treatment, adsorption is believed to be cost-effective and widely utilized. Activated carbon is a commonly used adsorbent but its production and regeneration is expensive. Development of low-cost adsorbents from industrial and agricultural wastes will be effective in that way, red mud as a low-cost adsorbent exhibit promising adsorption capacity. It is applied in both water and wastewater treatment. Red mud is effective for the removal of arsenic and phosphate from water.

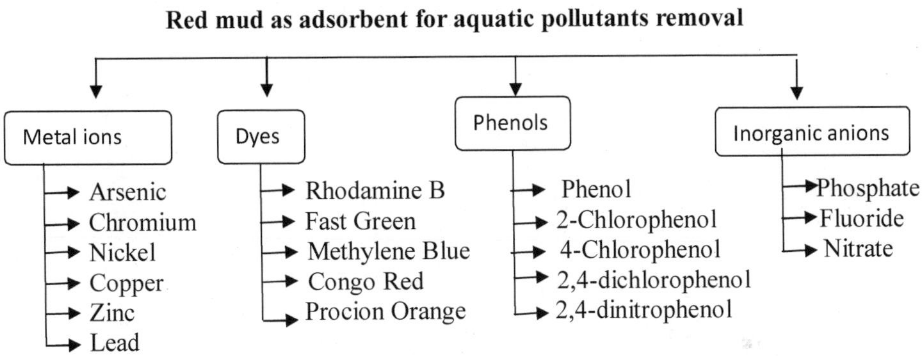

Figure 14.1: Red Mud as Adsorbent for Aquatic Pollutants Removal (Bhatnagar *et al.*, 2011).

Adsorption of trace elements/heavy metals and metalloids which are usually discharged from industrial processes. Most of heavy metals are toxic even at low concentration. Copper, zinc, lead, cadmium, and chromium are among the most heavily used metals in plating, mining and petroleum industries. For removal of heavy metals from wastewater systems, precipitation is the most widely used method. Due to the strong basic property of red mud with high pH, it can be used as an agent or adsorbent for surface precipitation and adsorption of heavy metals (Table 14.1). Gupta and Sharma (2002) also tested an ARM in batch and column operations for the removal of cadmium and zinc from aqueous solutions. The red mud was treated by heat-treated at 500 °C. The removal of $Cd^{2+}$and Zn^{2+}, Cd^{2+} was almost complete at low concentrations, while it was 60–65 per cent at higher concentrations at optimum pH values of 4.0 and 5.0 (Figure 14.2).

Phenol and substituted phenols are considered as priority pollutants. The discharge of effluents containing phenolic pollutants from various industries into natural water bodies is an ongoing and serious threat to human health and natural water quality. The ubiquitous nature of phenols, their toxicity even in trace amounts and the stricter environmental regulations make it necessary to develop processes for the removal of phenols from wastewaters. Red mud has also been explored as potential adsorbent for the removal of phenols from water and wastewater. The removal of 1-butanethiol from diesel oil at different concentrations and temperatures was investigated using red mud. Low concentration and high temperature favoured the percentage removal of 1-butanethiol from diesel oil.

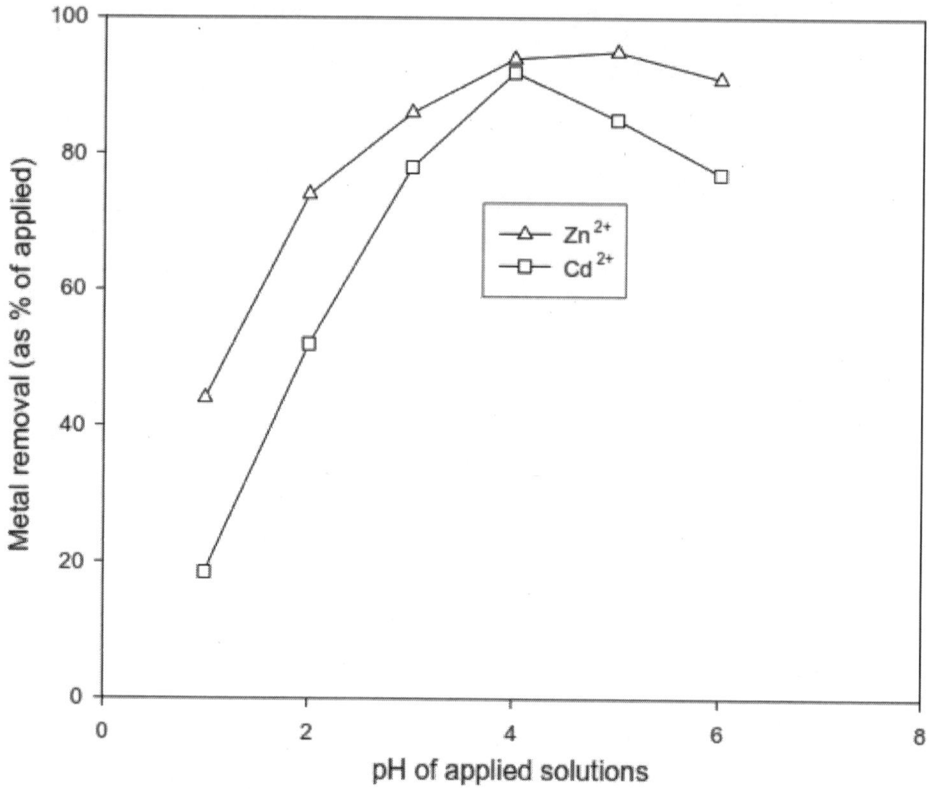

Figure 14.2: Effect of pH on the Adsorption of Metal Ions on Activated Red Mud (Wang, Ang and Tadé, 2008).

Red mud has also been used for the removal of phenol, 2-chlorophenol, 4-chlorophenol, and 2,4- dichlorophenol from wastewater (Wang *et al.*, 2008). The maximum adsorption of phenol and 2-chlorophenol occurred at pH 6.0, whereas the maximum adsorption of 4-chlorophenol and 2,4-dichlorophenol was achieved at pH 5.0 and 4.0, respectively. 2,4-Dichlorophenol and 4- chlorophenol were sorbed by the developed adsorbent by 94–97 per cent, while the removal of 2-chlorophenol and phenol was 50–81 per cent. Adsorption of mixtures of phenols was also tried by the researchers, and it was observed that 2,4-dichlorophenol had the maximum adsorption in comparison with the other phenols. The order of adsorption was 2,4-dichlorophenol > 4-chlorophenol > 2-chlorophenol > phenol, and the adsorption achieved was 97 per cent, 93 per cent, 80 per cent, and 51 per cent, respectively, with a concentration of 2 × 10–5 M of each phenol. The authors suggested that this sort of behaviour of red mud shows non-specificity with respect to phenols, and thus it could be used for the uptake of phenols individually or in their mixtures. The removal of phenol and its derivatives was up to 98 per cent by column experiments at a flow rate of 0.5 mL/min. The order of removal was 2,4-dichlorophenol > 4-chlorophenol > 2-chlorophenol >phenol, and the removal took place through a particle diffusion mechanism. Red mud was also used for the dynamic uptake of

2,4-dinitrophenol, which was sorbed by the developed adsorbent up to 95 per cent (Gupta *et al.*, 2006). The removal of this pollutant reached 96 per cent by column experiments at the flow rate of 0.5 mL/min. The adsorption process was found to be endothermic in nature.

Table 14.1: Adsorption Capacity of RM for the Removal of different Metals from Water

Adsorbent	Adsorbate	Amount adsorbed
Red mud	As(III)	8.86 μmol/g
Red mud	As(V)	6.86 μmol/g
Activated red mud	As(III)	11.80 μmol/g
Activated red mud	As(V)	12.57 μmol/g
Seawater-neutralized red mud (Bauxsol)	As(V)	6.08–14.43 μmol/g
Bauxsol-coated sand	As(V)	1.64–3.32 mg/g
Activated-Bauxsol-coated sand	As(V)	2.14 mg/g
$FeCl_3$-coated sand	As(V)	23.2–68.5 mg/g
CO_2-neutralized red mud (ANRM)	As(V)	55.55 mg/g
Activated red mud	Cr(VI)	30.74 mmol/g
Red mud	Ni(II)	13.69 mg/g
Red mud	Cu(II)	2.28 mg/g
Granular red mud	Cd(II)	38.2–52.1 mg/g
Red mud	Cu(II)	19.72 mg/g
Red mud	Zn(II)	12.59 mg/g
Red mud	Ni(II)	10.95 mg/g
Red mud	Cd(II)	10.57 mg/g
Treated red mud	Cd(II)	46.9–66.8 mg/g
Treated red mud	Cu(II)	35.2–75.2 mg/g
Treated red mud	Pb(II)	117.3–165.8 mg/g
Red mud	Pb(II)	64.79 mg/g
Red mud	Cr(VI)	35.66 mg/g
Red mud	Cd(II)	1.16×10^4 mol/g
Red mud	Zn(II)	2.22×10^4 mol/g
Red mud	Cd(II)	68 mg/g
Red mud	Zn(II)	133 mg/g
Calcined red mud	Copper	18.18–65.17 mg/g
Calcined red mud	Zinc	15.45–99.20 mg/g
Calcined red mud	Arsenic	18.83–27.51 mg/g
Neutralized red mud	Boron	30.12 mg/g

Source: Bhatnagar *et al.*, 2011.

Removal of Mn and As from drinking water by red mud (Pietrelli, Ippolito, Ferro, Dovi, *et al.*, 2019) has taken 2g of sorbent material, red mud (10 g L^{-1}) and initial concentrations of 2 ppm for Mn and of 100 ppb for As amount of Mn or As adsorbed (x) per gram of adsorbent (M), expressed as the ratio x/M, and percentage values of removal (Rem per cent) as a function of contact time. Test carried out with red mud show that both Mn and As could be removed almost completely that is 94 per cent and 99 per cent within 24 h of contact time Table 14.2.

Table 14.2: Adsorption of Mn and As on Red Mud 10 g $^{-1}$L at Time Interval

Time (min)	*Mn*		*As*	
	C(t) [ppb]	*Rem per cent Mn*	*C(t) [ppb]*	*Rem per cent As*
0	2000	0	100	0
5	1760	12	2.71	97.29
15	1750	12.5	1.71	98.29
30	1730	13.5	1.83	98.17
60	1725	10	0.5	99.46
120	1720	14	0.46	99.54
240	1480	26	0.17	99.83
720	800	55	-	-
1440	94	94	0.17	99.83

Source: Pietrelli *et al.*, 2019.

Table 14.3: Adsorption Capacity of RM for the Removal of Phenolic Pollutants from Water

Adsorbent	*Adsorbate*	*Amount Adsorbed*
RM	Phenol	**0.63-0.74mol/g**
RM	2-Chlorophenol	**0.72-0.79mol/g**
RM	4-chlorophenol	**0.78-0.82mol/g**
RM	2,4-dichlorophenol	**0.80-0.85mol/g**
Neutralized RM	Phenol	**4.12/g**
Acid-RM	Phenol	**8.16mg/g**

Source: Wang *et al.*, 2008.

Table 14.4: Adsorption Capacity of RM for the Removal of different Dyes from Water

Adsorbent	*Adsorbate*	*Amount Adsorbed*
RM	Rhodamine B	(1.01-1.16) x 10^{-5}mol/g
RM	Fast Green	(7.25-9.35)x 10^{-6}mol/g
RM	Methylene Blue	(4.35-5.23)x 10^{-5}mol/g
RM	Congo Red	4.05mg/g
RM	Acid Violet	1.37 mg/g
Acid-RM	Congo Red	7.08 mg/g

Source: Wang *et al.*, 2008.

Catalyst

Presently, numerous catalysts are available for industrial applications. Noble metals and oxides of metals are a highly preferable catalyst. Owing to high activation energies, noble metals are highly efficient catalysts in comparison with metal oxides. However, they are costly, toxic and exhibit low stability (Nguyen and Leclerc, 2007). An industrial waste-type catalyst consisting of metal oxide can be employed as a commercial catalyst for many reactions. Oxide of BR (Fe_2O_3) having potential alternative catalysts on a commercial scale. With minor treatments, BR-based catalysts are cheaper when compared with metal oxides and noble metals.

Properties such as iron content in form of ferric oxide (Fe_2O), high surface area, sintering resistance, resistance to poisoning and low cost makes it an attractive potential catalyst for many reactions. A constituent of BR-like iron content (Fe_2O_3), has a high surface area and is non-toxic in nature, which makes it a catalyst for many reactions (Wang *et al.*, 2008). Owing to its abundant availability, researchers utilized red mud as a catalyst for hydrogenation and liquefaction ((Wang *et al.*, 2008); (Yanik *et al.*, 2001)and in hydro dechlorination for destruction of chlorine-containing organics (Ordóñez *et al.*, 2001).

Gas Treatment

Treatment of Waste Gas

The production of considerable NOx, SOx, and H_2S in various industries leads to serious air pollution. The alkaline substance in red mud is abundant, which can purify and adsorb these gases. Red mud contains considerable alkaline substances, such as Al $(OH)_3$ and $CaCO_3$, which can react with an acidic gas. Most particles are amorphous and agglomerative, which form an agglomerative space. Therefore, RM has considerable pores, large specific surface area and excellent adsorption capacity (Sutar *et al.*, 2014).

$$SO_2\ (g) + Na_2O \rightarrow Na_2SO_3$$
$$4SO_2\ (g) + 4Na_2O \rightarrow 3Na_2SO_4 + Na_2S$$
$$4.5SO_2\ (g) + Al_2O_3 \rightarrow Al_2\ (SO_4)_3 + 1.5S$$
$$4SO_2\ (g) + 4CaO \rightarrow 3CaSO_4 + CaS$$

Metal Recovery

Iron is the major constituent of red mud followed by Al_2O_3, Na_2O, SiO_2, and TiO. Of these constituents, the recovery of titanium has the most potential value and, if successfully recovered, iron can be obtained as a value-added product, whereas alumina and alkali can be recycled in the Bayer's process stream. Depending on their amount in red mud, economic extraction of other metals can also be significant. A large number of processes are there for recovery of metals from red mud either exclusively or with combination used efficiently. The consumption of REEs is increasing due to an increase in demand for green technologies. The REEs are used in the production of permanent magnets, lamp phosphors, rechargeable batteries, catalysts and other applications. REEs are classified under the most critical raw materials group, with the highest supply risk. REEs are not rare; however, it is difficult to find their minerals in economic concentrations.

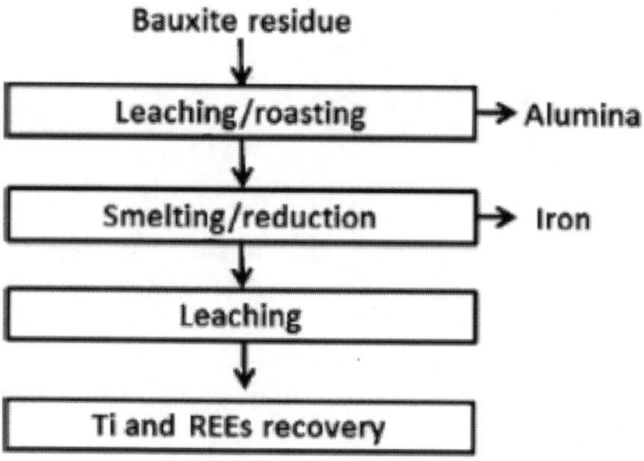

Figure 14.3: Process of Metal Extraction (Borra *et al.*, 2016).

Direct Leaching of Bauxite Residue

Leaching with organic or mineral acid such as Hydrochloric acid, sulphuric acid, nitric acid. The efficiency could be improved by leaching at higher temperatures.

Bioleaching

In bioleaching, metals are extracted with the aid of microorganisms. Two different kinds of micro-organisms are commonly used for bioleaching in extractive metallurgy bacteria and fungi. Bacteria are not suitable for bioleaching of bauxite residue due to the high pH and absence of energy sources (sulfur or reduced iron). However, fungi can survive in high pH and they excrete metabolites such as organic acids, amino acids, and proteins in the presence of an organic medium to form complexes with metal ions in bauxite residue.

Table 14.5: Metals Leaching using different Leaching Method (Qu *et al.*, 2015)

Leaching Mode	*Consumption*	*Metals Leaching Ratios from Red Mud (Per cent)*						
		Ga	*Ge*	*V*	*Sc*	*La*	*Eu*	*Yb*
Aspergillus niger	1t (sucrose)	31	33	19	30	16	23	44
H_2SO_4	110L	13	12	5	12	3	10	7
HCl	338L	17	14	10	9	5	11	22
Oxalic acid	232Kg	24	24	19	16	8	18	34
Citric acid	309Kg	33	30	17	27	16	25	41

Smelting

Red mud is treated in the blast or furnace in the presence of reducing whereby the iron oxides are reduced generating pig iron and a titanium-rich slag. The pig iron can be used for steel making and titanium from the slag can be recovered by leaching. This method quantitatively recovers the iron into a high-grade stream and separates it from other constituents.

Solid State Reduction

In the solid-state reduction process the mud is mixed with a reducing agent or contacted with a reducing gas at 800-1000°C, resulting in the formation of metallic iron, which could either be an input for a steel making furnace or perhaps a charge to a conventional blast-furnace. Although it involves less energy as well as capital cost compared to the blast furnace approach.

Reclamation of Soil

Red mud has a favorable environmental repair effect on the soil that has been contaminated by heavy metal elements. One of the explanations for the mechanism is that red mud can absorb heavy metal ions such as Cu^{2+} Ni^{2+}, Zn^{2+}, Pb^{2+}, Cd^{2+}, Cr^{6+}, Mn in the soil; the form of heavy metal ions changes from exchangeable ions into bonding oxides. Another mechanism is the precipitation reaction of carbonate in red mud with the heavy metal ions, and that causes these ions to deposit. In turn, the activity and reactivity of heavy metal ions in the soil are reduced, microbial activity and plant growth are promoted.

Red mud and compost are two inexpensive and effective materials available for the in situ treatment of soil contaminated with heavy metals. The efficiency of three different treatment systems, namely compost (C), red mud (R), and a mixture of compost and red mud (R and C), on the movability, ecological risk, and biological effective Cu, Zn, Cd, and Pb over time of incubation in agricultural soil showed significant changes in the pH, water extractable organic carbon (WEOC) and total organic carbon (TOC) after addition of these amendment materials to the target soil. The combination of red mud and compost mixture as amendment material (R and C) resulted in the largest reduction in efficiency of the available heavy metals.

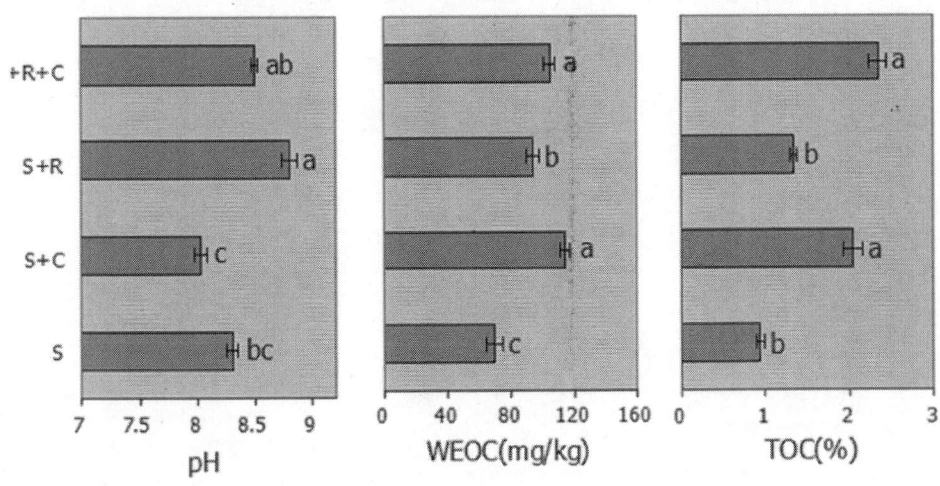

Figure 14.4: pH, WEOC and TOC of Soil with Soil Only (S), with Addition of Compost (C), Red Mud (R), Mixture of Compost and Red Mud (R and C) (Zhou *et al.*, 2017).

Red mud (RM) was used to remediate heavy metal-contaminated soils. Experiments with two different dosages of RM added to soils were carried out. It was found that soil pH increased 0.3 and 0.5 unit with the dosage of 3 and 5 per cent (wt per cent). At the dosage of 5 per cent, the highest stabilization efficiencies for Cd, Pb, Cu and Zn reached 67.95, 64.21, 43.73 and 63.73 per cent, respectively. The addition of RM transferred Cd from the exchangeable fraction to the residual fraction. In comparison with the control (no RM added), it reduced 24.38, 49.20, 19.42 and 8.89 per cent of Cd, Pb, Cu and Zn in wheat grains at the RM addition dosage of 5 per cent, respectively. The yield of wheat grains increased 17.81 and 24.66 per cent at the RM addition dosage of 3 and 5 per cent. The addition of RM did not change the soil bacterial community. These results indicate that RM has a great potential in stabilizing heavy metals in calcareous agricultural soils.(Zhou and Haynes 2011).

Other Applications

Photocatalytic Degradation

Red Mud-Based Geopolymers used for Photocatalytic Degradation of Methylene Blue towards Self-cleaning Green Construction Materials. Geopolymers are environmentally friendly alternatives to traditional cementitious materials as binder emerging application areas regarding geopolymers (GP) is utilizing them as "self-cleaning" green construction materials. Red mud as an aluminosilicate source and Tio_2- (5-10 per cent wt) provides mechanical strength, methylene blue solution dip in dark and UV light source for 6hrs. Red mud offer high adsorption/ degradation performance that can be improved further with the addition of TiO. They offer opportunities for rational design of self-cleaning green construction materials. Adsorption of MB took place in dark on all samples. Decrease in MB concentration was higher under UV-light indicating that both adsorption and degradation of MB occur under UV-light. RMGP displayed the highest degradation rate under UV-light (98.6 per cent) Degradation rates of MB after 6 hours (per cent) and crystallite sizes (nm) (Kaya *et al.*, 2019).

Table 14.6: Degradation Rates of MB after 6 Hours (per cent) and Crystallite Sizes (nm)

Sample	*Deg. Rate*		*Size of TiO_2*	*Size of Fe_2O_3*
	Dark	*UV*		
RMGP	63.6	98.6	-	34.9
5 per cent A RMGP	75.1	92	25.2	21.6
10 per cent A RMGP	69.7	94.6	41.8	27.3

Source: Kaya *et al.*, 2019.

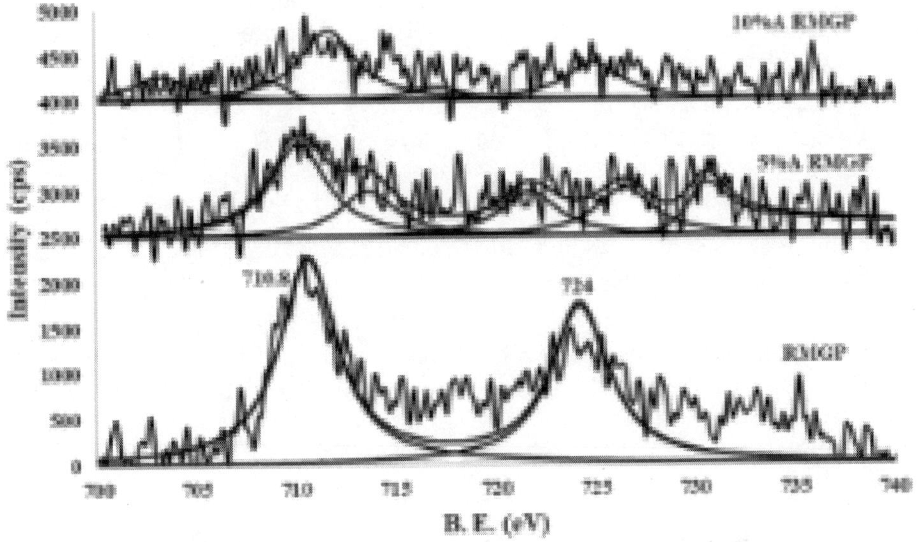

Figure 14.5: Fe Spectra of Red Mud Based Geopolymer (Kaya *et al.*, 2019).

REFERENCES

Palmer, SJ and Frost RL. "Characterisation of bauxite and seawater neutralised bauxite residue (Kaya *et al.*, 2019) using XRD and vibrational spectroscopic techniques." *Journal of materials science.* 2009, 44 (1): 55-63.

Paramguru, RK, PC Rath, and VN Misra. 2004. "Trends in red mud utilization–a review." *Mineral Processing and Extractive Metall. Rev.* 26 (1): 1-29.

Paredes, JR, S Ordóñez, A Vega, and FV Dýez. 2004. "Catalytic combustion of methane over red mud-based catalysts." *Applied Catalysis B: Environmental* 47 (1): 37-45.

Pietrelli, L., N. M. Ippolito, S. Ferro, V. G. Dovi, and M. Vocciante. 2019. "Removal of Mn and As from drinking water by red mud and pyrolusite." *J Environ Manage* 237: 526-533. doi: 10.1016/j.jenvman.2019.02.093.

Pietrelli, Loris, Nicolò Maria Ippolito, Sergio Ferro, Vincenzo Giorgio Dovì, and Marco Vocciante. 2019. "Removal of Mn and As from drinking water by red mud and pyrolusite." *Journal of environmental management* 237: 526-533.

Pohland, HH, and AJ Tielens. 1987. "Design and operation of non-decanting red mud ponds in Ludwigshafen." *Bauxite Tailings 'Red Mud'. The Jamaica Bauxite Institute*: 87-90.

Power, Greg, M Gräfe, and Craig Klauber. 2011. "Bauxite residue issues: I. Current management, disposal and storage practices." *Hydrometallurgy* 108 (1-2): 33-45.

Qu, Yang, Hui Li, Wenjie Tian, Xiaoqing Wang, Xuemeng Wang, Xiaohui Jia, Ben Shi, Gendi Song, and Yuan Tang. 2015. "Leaching of valuable metals from red mud via batch and continuous processes by using fungi." *Minerals Engineering* 81: 1-4.

Rai, S, KL Wasewar, DH Lataye, J Mukhopadhyay, and CK Yoo. 2013. "Feasibility of red mud neutralization with seawater using Taguchi's methodology." *International Journal of Environmental Science and Technology* 10 (2): 305-314.

Rai, Suchita, K Wasewar, J Mukhopadhyay, Chang Kyoo Yoo, and Hasan Uslu. 2012. "Neutralization and utilization of red mud for its better waste management." *World* 6: 5410.

Rai, Suchita, Kailas L Wasewar, Dilip H Lataye, Rajshekhar S Mishra, Suresh P Puttewar, Mukesh J Chaddha, P Mahindiran, and Jyoti Mukhopadhyay. 2012. "Neutralization of red mud with pickling waste liquor using Taguchi's design of experimental methodology." *Waste Management and Research* 30 (9): 922-930.

Rai, Suchita, KL Wasewar, and A Agnihotri. 2017. "Treatment of alumina refinery waste (red mud) through neutralization techniques: A review." *Waste Management and Research* 35 (6): 563-580.

Robinsky, EI. 1987. "Current status of sloped thickened tailing disposal system." *Bauxite Tailings 'Red Mud'. The Jamaica Bauxite Institute*: 91-100.

Sahu, MK, and RK Patel. 2016. "Methods for Utilization of Red Mud and Its Management." In *Environmental materials and waste*, 485-524. Elsevier.

Þayan, Enes, and Mahmut Bayramoðlu. 2004. "Statistical modeling and optimization of ultrasound-assisted sulfuric acid leaching of TiO2 from red mud." *Hydrometallurgy* 71 (3-4): 397-401.

Singh, Suresh Prasad, S Samantasinghar, and D Sindhuja. 2017. "Stabilization of Red Mud Using Ground Granulated Blast Furnace Slag by Geopolymerization Technique." International Congress and Exhibition" Sustainable Civil Infrastructures: Innovative Infrastructure Geotechnology".

Sutar, Harekrushna, Subash Chandra Mishra, Santosh Kumar Sahoo, and Himanshu Sekhar Maharana. 2014. "Progress of red mud utilization: An overview." *Chemical Science International Journal*: 255-279.

Szirmai, Endre, Sandor Babusek, Gezz Balogh, Atilla Nedves, Gyula Horvath, Zoltan Lebenyi, and James Pinter. 1991. Method for the multistage, waste-free processing of red mud to recover basic materials of chemical industry. Google Patents.

Tang, WC, Z Wang, SW Donne, M Forghani, and Y Liu. 2019. "Influence of Red Mud on Mechanical and Durability Performance of Self-compacting Concrete." *Journal of hazardous materials*: 120802.

Thakur, RS, and SN Das. 1994. "International series on environment in red mud—analysis and utilization." *PID and Wiley Eastern Ltd., New Delhi, India*.

Verma, Ajay S, Narendra M Suri, and Suman Kant. 2017. "Applications of bauxite residue: A mini-review." *Waste Management and Research* 35 (10): 999-1012.

Wang, Li, Ning Sun, Honghu Tang, and Wei Sun. 2019. "A Review on Comprehensive Utilization of Red Mud and Prospect Analysis." *Minerals* 9 (6): 362.

Wang, S., H. M. Ang, and M. O. Tade. 2008. "Novel applications of red mud as coagulant, adsorbent and catalyst for environmentally benign processes." *Chemosphere* 72 (11): 1621-1635. doi: 10.1016/j.chemosphere.2008.05.013.

Wang, Shaobin, HM Ang, and MO Tadé. 2008. "Novel applications of red mud as coagulant, adsorbent and catalyst for environmentally benign processes." *Chemosphere* 72 (11): 1621-1635.

Wang, Ying, Yange Yu, Haiyan Li, and Chanchan Shen. 2016. "Comparison study of phosphorus adsorption on different waste solids: Fly ash, red mud and ferric–alum water treatment residues." *Journal of Environmental Sciences* 50: 79-86.

Yang, Xiaojie, Jianfeng Zhao, Haoxin Li, Piqi Zhao, and Qin Chen. 2017. "Recycling red mud from the production of aluminium as a red cement-based mortar." *Waste Management and Research* 35 (5): 500-507.

Yanik, Jale, Md Azhar Uddin, and Yusaku Sakata. 2001. "The effect of red mud on the liquefaction of waste plastics in heavy vacuum gas oil." *Energy and fuels* 15 (1): 163-169.

Zhang, Jishi, and Lihua Zang. 2016. "Enhancement of biohydrogen production from brewers' spent grain by calcined-red mud pretreatment." *Bioresource technology* 209: 73-79.

Zhou, Rui, Xiaochen Liu, Lin Luo, Yaoyu Zhou, Jianhong Wei, Anwei Chen, Lin Tang, Haipeng Wu, Yaocheng Deng, and Fengfeng Zhang. 2017. "Remediation of Cu, Pb, Zn and Cd-contaminated agricultural soil using a combined red mud and compost amendment." *International Biodeterioration and Biodegradation* 118: 73-81.

Zhou, Ya-Feng, and Richard John Haynes. 2011. "Removal of Pb (II), Cr (III) and Cr (VI) from aqueous solutions using alum-derived water treatment sludge." *Water, Air, and Soil Pollution* 215 (1-4): 631-643

Chapter 15

Biotechnology, Genetic Improvement in Relation to Climate Change and Food Security

Manisha Kumari[1], Piyush Choudhary[2] and Amit Godara[3]*

[1]Department of Genetics and Plant Breeding, [2]Department of Agronomy, Rajasthan College of Agriculture, Maharana Pratap University of Agriculture and Technology, Udaipur – 313 001, Rajasthan

[3]The Centre for Global Food and Resources, University of Adelaide, Adelaide, South Australia - 5005

Several climate change related factors including temperature increases, changes in rainfall patterns and outbreak of pests and diseases negatively affect agricultural productivity and food security. Climate change effects significantly increase production risk and rural vulnerability, particularly in regions that already suffer from chronic soil and water resource scarcity or high exposure to climatic extremes, such as droughts and flooding. The potential to improve the livelihoods of resource-poor farmers is a strong incentive to meet the challenge. Climate change is having increasing impacts on food security globally. With 925 million citizens expected to be undernourished by 2020, the contribution of biotechnology to meeting world food security needs is discussed. Applications of biotechnology including marker aided selection, genomics, genetic modifications and emergent novel technologies of relevance to drought resistance and salt tolerance to meet the needs of an increasing population under stressed environmental conditions are described.

Keywords: Climate change, Food security, Biotechnology.

Introduction

Climate change is emerging as a key threat to global and national food security in the coming decades. Climate change related factors including rising temperature, variable patterns of precipitation and outbreak of pests and diseases, negatively affect agricultural productivity and food security. Growing world's population, increasing demand for healthy and culturally accepted food, soaring food prices and global environmental change challenges are negatively affecting food security globally. Climate change continually challenges our ability to feed the increasing global population. Agricultural production is extremely vulnerable to climate change, and its disruption will make it difficult to support the estimated world population in 2050. According to United Nations (UN) Food and Agricultural Organization (FAO) estimates, the world population will exceed 9.7 billion in 2050 (FAO, 2018), and approximately 57 per cent of this population (5.5 billion people) will occupy Asia. The effects of climate change on agriculture may depend not only on changing climate conditions, but also on the agricultural sector's ability to adapt through changes in technology and demand for food. Significant proportions of the growing populations in developing countries derive their livelihoods from agriculture and are, therefore, vulnerable to climate change effects. While several factors are responsible, high food prices are driving millions of people into food insecurity, worsening the conditions for many who are already food-insecure and threatening long-term global food security. The task of reducing the number of hungry people (as per Millennium Development Goals) will require both regional and global efforts among which biotechnology adoption could play a crucial role. To cope with the global food and energy crises as well as environmental problems and make human life inclusive, safe, resilient, and sustainable, it is essential to strengthen our ability to develop new improved cultivars using modern agricultural biotechnology. Plant biotechnology can be used as a tool to maximize plant productivity by introducing stress tolerance genes and key metabolic genes from will germplasm into environmentally adapted cultivars (Kwak, 2019).

Effect of Climate Change on Food Security

Food security and global environmental change are interrelated and there is growing concern that while global environmental change complicates achievement of food security, increasing demand for food further degrades the environment. Climate change may affect food systems in several ways, ranging from direct effects on crop and livestock production to changes in markets and food prices. Potential effects of changes in climate on crop yields have been assessed by researchers. Climate change and other global environmental changes such as changes in water availability, land cover, altered nitrogen availability and cycling (all strongly influenced by human activities), have increased concerns about achieving food security especially for poor people.

Climate change and its potential implications for agricultural water use create increasing uncertainty for planning. It can significantly increase production risk and rural vulnerability, particularly in regions that already suffer from chronic soil and water resource scarcity or high exposure to climatic extremes, such as droughts and flooding. All of these can lead to poverty and hunger. The effects of climate change on agriculture may depend not only on changing climate conditions, but also on the agricultural sector's ability to adapt through changes in technology and demand for food, coupled with management of water availability, soil quality, and crop selection. Adapting to climate change will require on-going review of its impact on population growth, land and water use, the effectiveness of water demand-and-supply programs, and the capacity of emerging technologies to deal with challenges. Changes in precipitation can affect soil erosion rates and soil moisture, both of which are important for crop yields. Agriculture in developing countries is expected to be more vulnerable to climate change than industrialized countries due to limited resources for adaptation.

Climate change will add a further dimension to the challenge of ensuring food security globally. The damage inflicted on the environment is becoming increasingly evident. This includes arable lands lost to erosion, salinity, desertification, deforestation, loss of biodiversity, and emerging water scarcity. This situation will be significantly exacerbated by climate change, as well as by the increase in extreme weather events and climate variability. Thus, food insecurity is likely to increase under climate change, unless early warning systems and biotechnologies are used more effectively. Climate change and its interaction with the above drivers will create demand for new food products that improve mitigation of the environmental impact of animal and food production. Biotechnology application is key to creating new food products. This requires investments in basic and applied research in order to respond to climate change effects. Research to alter animal and plant genetics to respond to changing climates, natural resource redistributions, and disease resistance are also required

The Role of Biotechnology in Food Security

Biotechnology is the use of living organisms to enhance crops, fuels, medical treatments and a host of other tools that can help humans. Modern biotechnology manipulates the genes of organisms and inserts them into other organisms to acquire the desired trait. Biotechnology contributes to poverty alleviation by increasing crop productivity and income generation, particularly for resource-poor farmers, and brings about a safer environment and more sustainable agricultural development. Advances in biotechnology can lead to cutting-edge technologies in crop agriculture in the decades ahead. Plant science and molecular biotechnology can solve this problem through an improved knowledge of host-pathogen interactions, the development of new diagnostic tools, and the use of genetic modification to create resistant crop-plant varieties.

The biotechnology tools that may be used in research are tissue culture and molecular techniques. The molecular techniques are DNA fingerprinting or characterization, marker assisted selection (MAS), molecular diagnostics and genetic engineering/transformation/genetic modification. The use of "Genetically Modified Organisms" (GMOs) has the potential to offer real benefit in agriculture, food quality, food nutrition and health. GMOs are plants or animals created through the gene splicing techniques of biotechnology.

Emergent Technologies for Regulating Gene Expression in Food Crops

Amongst the areas where new technology are likely to influencing the use and growth of food crops in response to climate change, three approaches stand out:

Gene Silencing

It is a means of down regulating (or 'turning off') particular genes by over expression of RNA sequences, known as RNAi, preventing functional expression of a gene. This RNAi technology can be applied using DNA from sexually compatible wild relatives of crop plants, as in Innate® potatoes, which should make gaining regulatory acceptance easier.

Gene Editing

It is a means of making precision, directed changes in genomes at as fine a scale level as one, or a few nucleotides (Ledford, 2015a). Two alternative systems currently provide state of the art protocols for achieving these small scale genomic changes, using clustered regularly interspaced short palindromic repeats (CRISPR) and the CAS9 nuclease, or alternatively, transcriptional activator-like effectors nucleases (TALENS).

RNA Spraying Technology

This technology topically applies specific synthetic RNA to the surfaces, *e.g.* leaves of plants to control plant responses or stimulate pathogen resistance. This technology removes the need to use genetic modification in such applications, as no change to the plant genome takes place.

Attitudes, Needs and the Future

Biotechnology tools are and will continue to make positive contributions to enhancing food security during climate change, alongside a range of other means to ensure food availability, access to food for all, efficient utilization of food resources and a stable global food commodity trading system. The extent to which biotechnology will help to achieve this is dependent on several factors, including the rate of technological development, governmental and public acceptance of novel biotechnologies, and the costs of climate change effected food crops to consumers. For the increasing numbers of undernourished people, as the global population grows towards 9.6 billion by 2050 (Federoff, 2015), particularly in the developing nations, choices about how a food crop has been produced are likely to be an unaffordable luxury.

REFERENCES

FAO (2018). The future of food and agriculture: alternative pathways to 2050. Food and Agriculture Organization of the United Nations, Rome.

Federoff, N.V. (2015). Food in a future of 10 billion. *Agriculture and Food Security.* 4 (11.)

Kim, H.S. and Kwak, S.S. (2020). Crop biotechnology for sustainable agriculture in the face of climate crisis. *Plant Biotechnology Reports.* 14: 139–141.

Kwak SS (2019). Biotechnology of the sweet potato: ensuring global food and nutrition security in the face of climate change. Plant Cell Rep 38: 1361–1363

Ledford, H. (2015a). CRISPR, the disruptor. *Nature.* 522 20-24.

Quaye W., Yawson R.M., Ayeh, E.S. and Yawson I (2012). Climate Change and Food Security: The Role of Biotechnology. *African Journal of Biotechnology.* 12(5): 6355-6364.

Chapter 16

An Evaluation Study on Sustainability of TN-IAMWARM Project Climate Smart Agricultural Interventions in Tamil Nadu

S. Kanaka and N. Kavitha

[1]Assistant Professor, Agricultural Economics, Tamil Nadu Agricultural University, Coimbatore, Tamil Nadu
[2]Senior Project Adviser, Department of Ocean Engineering, IIT, Chennai, Tamil Nadu

An evaluation has been conceptualized to study the present status and sustainability of climate smart agricultural interventions adopted in the different sub basins and the respective tanks falling in Tamil Nadu State pertaining to the TN -IAMWARM project period. In order to objectively assess the status and sustainability, along with the 125 tanks from 41 sub basins which were served under IAMWARM-I, 36 tanks from 28 sub basins were selected as 'control' which were not served by the project. The present paper proposed an approach to the evaluation of farm sustainability. The issue of the diversification and completely linked with it, the issue of gap area coverage in the Ayacut area also to be discussed. The extent of increase in the Horticulture crops found in the project area. The area shifted and sustained to diversified crop. The extent of adoption of INM and IPM practices. 10 to 20 per cent shift from paddy to other commercial crops, 30 to 40 per cent adopted of INM and IPM practices. 10 to 20 per cent increase in productivity. 10 to 20 per cent increase in farm income.

Keywords: *TN-IAMWARM, Sustainability, Climate smart agriculture, System of rice intensification, Crop diversification, Convergence.*

Introduction About the TN-IAMWARM Project of Tamil Nadu

In a predominantly Agricultural State like Tamil Nadu, there is need for intensifying efforts to improve productivity, and sustainable farm income. Long-term growth in agriculture depends in large part on increasing the efficiency and productivity of use of water. A concomitant need in this area is the strengthening and integrating of institutional structures which can help small and marginal farmers' access responsive irrigation management and improved agriculture practices. Faster growth in agriculture is central to sustainable development and poverty reduction in Tamil Nadu. Although agriculture accounts for only 15.7 per cent of total GSDP, farm income accounts for about half of household income for 35 million people (56 per cent of the state's population) who live in rural areas. Much of this rural population is poor, with estimates ranging from 20.6 per cent to 31 per cent of rural population. For the poorest rural quintile (approximately 1.5 million households, or 7.5 million people), more than three-quarters of income is derived from agriculture, with agricultural wage labour alone accounting for half of household income. Given the importance of agriculture in the incomes of the poor in Tamil Nadu, growth in agriculture could further reduce rural poverty through higher yields to agricultural producers, higher real wages to agricultural labourers, and increased income and employment opportunities with forward and backward links to the rural non-farm sector. The Tamil Nadu has been engaged with this critical challenge evolving various responses over time. An Expert Committee on "Development and Management of Water Resources" constituted as per G.O. Ms. No. 332PW(R1) Dept. dated 6.7.2000 observed that bringing additional area into cultivation is remote but the challenge is how best to bridge the gap in cultivation by reducing demand, by effective water management and by adoption of modern agricultural techniques (Micro Irrigation *etc.*) The Committee recommended Integrated Water Resources Management and convergence of various Departments for development and management of water resources in Tamil Nadu.

With Agriculture sector facing major constraints due to dilapidated irrigation infrastructure, coupled with water scarcity (both quantity and quality) and growing demands from industry and domestic users, long term growth in agriculture and rural income depends in large part on increasing efficiency and effectiveness in the use of water. Concomitantly increased agricultural diversification and private investments in higher value processing are likely to generate new rural non-farm employment opportunities and raise rural incomes. Increased availability of water and greater efficiency of water use through widespread adoption of drip and sprinkler irrigation could enable cultivation of crops over larger area, year round, providing employment in agricultural production and processing, benefiting the rural poor. It is important to ensure that the ultimate outcome of irrigated agriculture is food security and improved farm incomes. The four major areas of focus present themselves:

1. Improved infrastructure
2. Crop diversification
3. New technologies for productivity
4. Market participation

It is, with this back ground, the Irrigated Agriculture Modernization and Water-Bodies Restoration and Management (IAMWARM) project is prepared converging all Line Departments.

Tamil Nadu Irrigated Agriculture Modernization and Water-Bodies Restoration and Management (TN-IAMWARM) is a multidisciplinary project funded by the World Bank and implemented by the Water Resources Organization (WRO), PWD and Government of the Indian state of Tamil Nadu as the Nodal Agencies. Tamil Nadu Agriculture University (TNAU) concentrates on the upscaling of water saving technologies in rice and major garden land crops. The total out lay for TNAU is Rs.88.90 crores.

The focal technologies are systems of rice intensification (Rajarajan 1000), improved production technologies (IPT) on garden, cotton, pulses, maize, sunflower, groundnut, and gingelly. Promotion of new crops such as thorn less bamboo and cocoa intercropping in coconut is also programmed. Precision farming with drip fertigation is also demonstrated with sugarcane, banana, vegetables, tapioca, coconut and flowers. The concept of an organic farming seed village is promoted in the sub basins. E-velanmai technology will be tested on a pilot basis. Labour saving implements such as a tractor drawn seed drill, maize husker cum sheller and sunflower thresher introduced. A mini dhal mill is introduced. Implements such as a coconut shredder introduced under waste recycling. Vermi bag – a cost-effective portable unit demonstrated in the project.

Project Specific Climate Smart Agriculture Interventional Objectives

- ☆ Working towards large scale adoption of specific technologies such as SRI in Paddy, improved production technologies in oilseeds, pulses and cotton to increase the productivity.
- ☆ Large scale adoption of Drip Fertigation/Precision Farming and Sprinkler irrigation technologies in field crops, Commercial crops and horticultural crops to impart application efficiency.
- ☆ Popularization of crop diversification options in select crops such as maize, fruits and vegetables in the sub basin commands.
- ☆ Popularization of labour saving implements in rice, groundnut, cotton and maize to reduce the cost of production.
- ☆ Testing and large scale adoption of technologies developed on-station to farm conditions.
- ☆ Promotion of Seed Village Concept (SVC) to produce and supply quality seed and seedlings for diversified agricultural practices.
- ☆ Provide information on Price Forecasting and Market Intelligence to Water Users Association to develop more market access.
- ☆ Provide necessary trainings through various centre/research stations of Tamil Nadu Agricultural University to the farmers and other stakeholders for capacity building.

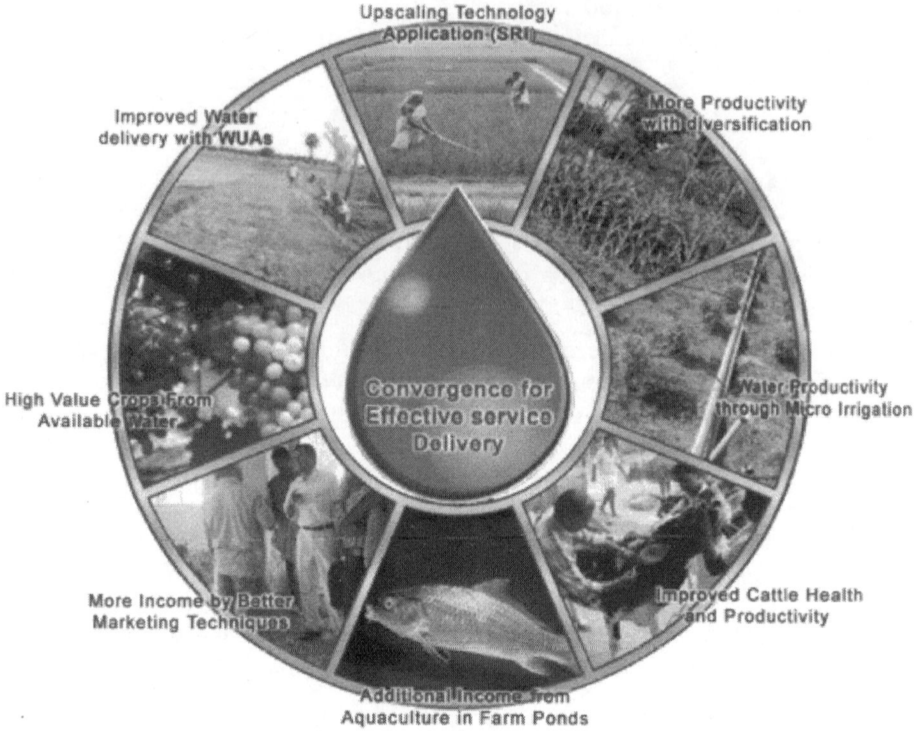

Figure 16.1: TN-IAMWARM Project Convergence for Effective Service Delivery.

The project covers 6.17 Lakh Hectares of Ayacut areas in four phases. 55 sub basins were covered in year 2013 with an outlay of Rs.2547 crores.

Phase	*Year*	*Sub Basins*
Phase I	2007-08	09
Phase II	2008-09	16
Phase III	2009-10	25
Phase IV	2011-12	04

The Motto of IAMWARM PROJECT "is *More Money per Litre of Water*" should be engraved in the mindset of the implementing officers. Implementation Officers may be given orientation training on the IAMWARM concept so as to ensure effective implementation. Focus on how precision farming and improved production technologies enhance the productivity of the plant increased production and therefore higher income (Figure 16.2).

Reasons for this Evaluation of Sustainability of Climate Smart Agricultural Technology

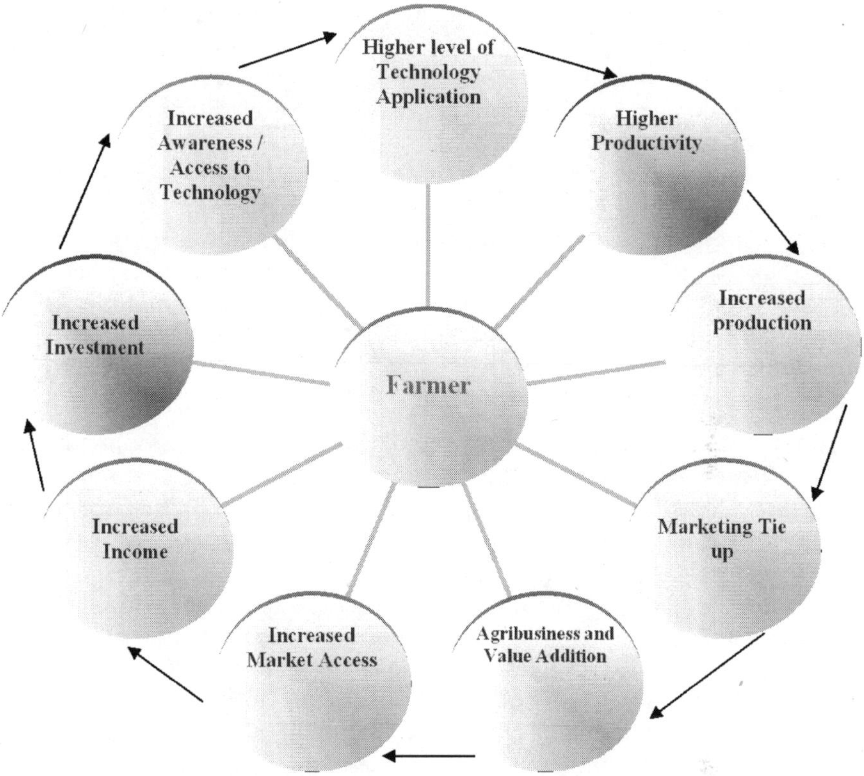

Figure 16.2: Technology – Productivity – Profitability (TPP) Cycle.

Studies

Water is becoming increasingly scarce resource and limiting agricultural development in many developing and developed economies across the world. It is estimated that around 50 per cent of the increase in demand for water by the year 2025 can be met by increasing the effective use of irrigation water [1]. In India, almost all the easily possible and economically viable irrigation water potential has already been developed, but the demand for water for different sectors has been growing continuously [2,3]. Moreover, the water use efficiency in the agricultural sector, which still consumes over 80 per-cent of water, is only in the range of 30 - 40 per cent, indicating that there is considerable scope for improving the water use efficiency.

It is well known fact that tanks are small reservoirs that are generally constructed in villages for multiple purposes such as domestic, irrigation, livelihood, groundwater recharge; these tanks are considered as one of the biggest assets of villages and are common to all villagers. Tank Irrigation system is a cost effective irrigation system since most of the farmers in the state are marginal farmers. Tanks are classified as system tank and non-system tank based on the inflow, PWD tanks and panchayat tanks based on the ayacut area. A tank has its catchment area, feeder channel, embankment, sluices, weirs, field channels and ayacut area. Most of the tanks existing in the state of Tamil Nadu are century years old. Due to inadequate rainfall, encroachment on tank bed, poor maintenance of tank components the tank irrigation system keeps on decreasing. The share of tank irrigation in Tamil Nadu declines to 19 per cent and the share of well irrigation increased to 50 per cent [4]. There is an urge to rehabilitate these tanks and bring their condition to an improved state. By knowing the significance of tanks the state Government is trying to rejuvenate these tanks under various schemes, many international agencies are also funding for tank rejuvenation in south Asia especially in south India. One such project is TN-IAMWARM, Tamil Nadu Irrigated Agriculture Modernization and Water-Bodies Restoration and Management (IAMWARM) is a Multidisciplinary Project funded by the World Bank.

Tamil Nadu Irrigated Agriculture Modernization and Water Bodies Restoration and Management (TN-IAMWARM) Project primarily aims at rehabilitating tanks and other systems in the state to improve their water storage capacities and utilize water efficiently and economically to achieve the iconic objective of getting more income per drop of water. The project design provides for a wide range of interventions for rehabilitating the tank systems and improving agricultural productivity and production. The proposed interventions were implemented in a phased manner; spread over four phases, covering 5,012 tanks from 61 sub-basins in the state. Implementation of the project was started during the year 2007-08 and proposes to complete by September, 2014. A Multi-Disciplinary Project Unit (MDPU) an apex body for implementation, consisting of eight line departments of the State Government and Tamil Nadu Agricultural University (TNAU), has co-ordinate the activities of all the implementing agencies. Each implementing agency had a clearly articulated and targeted action plan to be executed at the sub-basin level.

About the Evaluation Exercise

An evaluation has been conceptualized to study the present status and sustainability of climate smart agricultural interventions adopted in the different sub basins and the respective tanks falling in Tamil Nadu State pertaining to the TN -IAMWARM project period. In order to objectively assess the status and sustainability, along with the 125 tanks from 41 sub basins which were served under IAMWARM-I, 36 tanks from 28 sub basins were selected as 'control' which were not served by the project.

Objective

To study the sustainability of the IAMWARM successful interventions demonstrated sub basin tank areas/beneficiary farmers field of the sub basin.

Methodology

The Data Collection Tool

A structured interview schedule prepared from the researcher consisting of items which were intended to capture the present status and compare IAMWARM modernized tank with control tanks with the relevance to their performance was used as data collection tool. All the Tamil Nadu Agricultural university centres led the survey by direct data collection. Both primary and secondary data were collected for this study. Secondary data like total ayacut (command) area, agricultural cultivated area, rainfall and other hydrological details were collected from different Tamil Nadu government web portals and records. Water storage in the tank over the period and area irrigated in the command –net and gross, source wise and crop wise details were also collected from the village administrative officer. Primary data was collected through pre-tested interview schedule, pay personal visits to the tank villages. Percentage analysis, average, correlation and trend analysis are the tools of analysis followed in this study. The below map shows the entire study area of sub basin wise and district wise in the IAMWARM project.

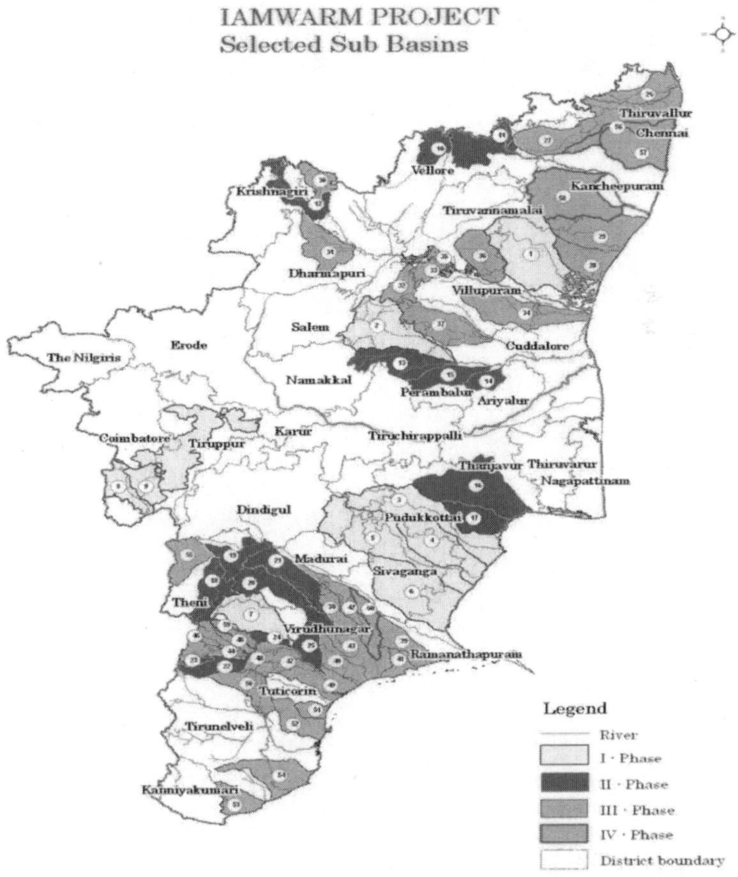

Figure 16.3: TN-IAMWARM District-wise 61-Sub Basins.

The Respondents of the Study

Twenty five farmers in each of the selected project and control tanks were selected randomly those are either beneficiary or non beneficiary of IAMWARM. Primarily, the farmers were asked to spell the name of the fellow farmers who were involved in IAMWARM interventions. The process went on until the required number of respondents arrived at. Totally 3125 respondents were contacted which comprises of 125 project tanks from the tanks which were covered under IAMWARM and 900 respondents from the 36 control tanks.

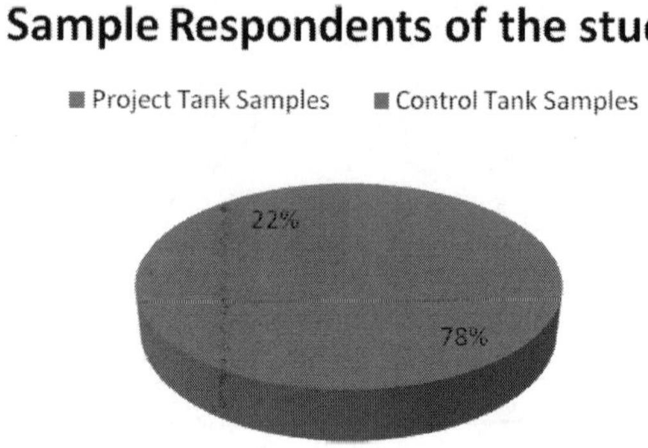

Trend Analysis

When a series of measurements of a process are treated as a time series, then the application of trend estimation can be used to make and justify statements about trends in the data. This model can then be used to explain the behaviour of a measurement. In particular, it may be useful to determine if measurements exhibit an increasing or decreasing trend, which is statistically distinguished from random behaviour. In this study the rainfall, area and production trend of the state was analysed.

Results and Discussions

Collected tank data were analyzed and are discussed below

General Characters of IAMWARM Tanks and Control Tanks

Tanks have been influenced by three important factors *viz.*, hydrological factor, socio-economic behavior of farmers and administrative factors.

Table 16.1: General Characters of IAMWARM Tanks and Control Tanks

IAMWARM Tank				N=125		Control Tank				N=36	
System	Rainfed	Single Village	Multi Village	Single Crop	Double Crop	System	Rainfed	Single Village	Multi Village	Single Crop	Double Crop
44.4 per cent	63.6 per cent	89.6 per cent	10.4 per cent	58 per cent	42 per cent	20 per cent	80 per cent	100 per cent	-Nil-	98 per cent	02 per cent

Table 16.1 shows that the tank and village sharing and single or double crop particulars collected from the project and control tanks. The IAMWARM tanks nearly 45 per cent of the tanks are system tank it is fed with water from rivers and runoffs through diversion weirs, feeder channels and surface flow. They may be linked as cascades where water from one upper tank flows into the lower tank. Non system tank are 64 per cent are these tanks, which receive their water intake from the rain. The control tanks almost 80 per cent are rainfed or non system tanks. The village sharing system is only 10 per cent is found in the IAMWARM tanks. Double crop has cultivated 42 per cent in the IAMWARM tank villages because of the full tank level achieved by the modernization of the tank. In the case of control tank only two per cent has cultivated.

Table 16.2: Area Irrigated and Average Rainfall in the Commend Area

Year	*Project tank*			*Control Tank*			*Avg. Rainfall*
	FI	*PI*	*EG*	*FI*	*PI*	*EG*	
2014-15	46.12 (70.44)	10.38 (15.85)	8.97 (13.70)	19.62 (30.84)	21.55 (33.88)	22.43 (35.26)	78.84
2015-16	45.87 (70.06)	11.8 (18.02)	7.8 (11.91)	18.34 (28.83)	22.18 (34.87)	23.08 (36.28)	78.62
2016-17	19.99 (30.53)	20.26 (30.94)	25.22 (38.52)	1.5 (2.35)	2.9 (4.55)	59.2 (93.08)	32.46

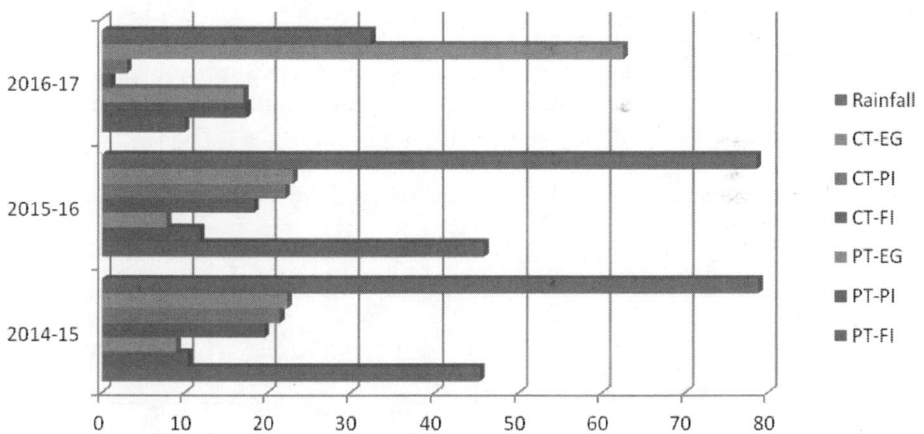

Figure 14.4: Average Rainfall and Area Irrigated in the Commend Area

(FI: Fully Irrigated Area; PI: Partially Irrigated Area; EG: Existing Gap).

The average rainfall during the sustainability study period was more than 75 per cent except the last year received the 32 per cent. The table clearly shows the increased rainfall given the appreciable cultivation in the fully irrigated areas of the sub basins. The less rainfall increases the existing gap area from 36 per cent to 93 per cent in the without project areas. Whereas the project area the existing gap area has been less than 40 per cent because of the water saving technologies followed by the sub basin farmers.

Table 16.3: Crop Performance in Sample Tank Ayacuts in the Current Year

Crop Performance	*Project Tanks (PT)*		*Control Tanks (CT)*	
	No	*Per cent*	*No*	*Per cent*
Successful - 100 per cent area	2	1.6	0	0
75 per cent area	5	4.0	0	0
50 per cent area	14	11.2	2	5.13
25 per cent area	56	44.8	2	5.13
Complete failure	48	38.4	35	89.74
Total	**125**	**100**	**39**	**100**

It is observed that out of 125 project tanks, the crops were successful in 2 (1.6 per cent) ayacuts while in 48 tanks (38.4 per cent) the crops failed completely. In the remaining tank ayacuts the area under successful cropping varies from 25 (44.8 per cent) to 75 per cent. Out of 39 control tank ayacuts, the cropping was not successful in the tanks whereas in the case 35 tanks (89.74 per cent) the crops failed completely. In the remaining tank ayacuts the area under successful cropping varies from 25 to 50 per cent. When comparison between project tanks and control tanks on crop failure is made, the project tanks are found to be better at least 45 per cent could tolerate the drought condition and cultivated some crops – complete failure was in 38.4 per cent of the tanks as compared to 89.74 per cent in control tanks is much heavier loss. This may be due to the impact of the project in augmenting water supply in the tanks and better groundwater recharge in the project tank ayacuts.

Table 16.4: General Particulars of the Farmers (N= 4025)

Sl.No.	*Land Owned by the Respondents*	*Number of Respondents*	*Percentage*
1.	Below 0.5 ha	1067	26.51
2.	0.5-1 ha	1563	38.83
3.	1-2 ha	870	21.61
4.	Above 2 ha	525	13.04

The above table indicated that majority of the respondents had lands in the category of 0.5-1 ha is 38.83 per cent and an appreciable number of them had land above 2 ha also.

Average Total Land Cultivated in Acres

An assessment of the average land area cultivated by the respondents before and after the project was increased 0.285 ha (0.71 acres) the details presented in the following table.

Table 16.5: Average Total Land Cultivated in Acres

Tank Type	*Before Project (ha)*	*After Project (ha)*	*Per cent Change*
Project Tank	0.996	1.41	41.56
Control Tank	0.878	0.89	1.37

The average land area irrespective of the crop has been found to be marginally improved after the IAMWARM interventions and also the increased water supply of Head and Tail area in the project tank ayacut. Typically, most of the sub basins treated in IAMWARM Phase I in Tamil Nadu has rice as major crop and the respondents told that they could increase their area in rice only that too under SRI cultivation and the vegetable growers under precision farming is also gradually increased their cultivated area because of the less water requirement. In the case of control tank the area also increased 0.062 ha. This meager area increase we could found that area is the impact of the IAMWARM interventions such as SRI and precision farming in vegetables.

Average Additional Area brought under Cultivation

The respondents were asked about the additional area that they brought in both the IAMWARM treated tanks and control tanks and the results are presented below in the following figure.

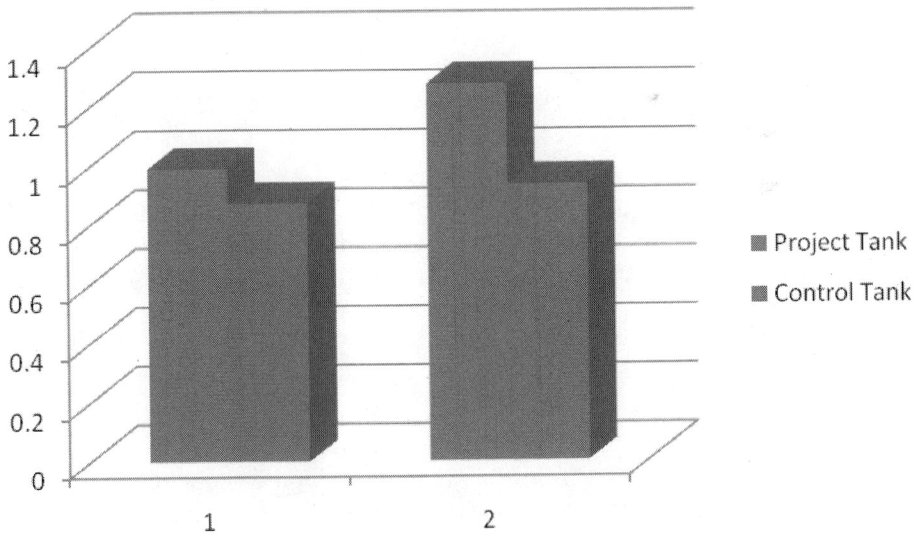

Figure 16.6: Additional Area Brought Under Cultivation (in ha).

It is found from the above table that in the tank cultivated areas treated under IAMWARM, there was a slight increase in the additional area brought under cultivation which surpasses the additional area brought under cultivation in control tanks. This might be due to the improved yield that farmers experienced in rice cultivation through SRI. In Pungankulam, One Mr. Ramachandran, who was a beneficiary of IAMWARM I project expressed that he could able to get 12 bags extra due to the shift in rice cultivation practice from conventional to line planting. He told that they did not go for all components of SRI, except line planting. The line planting in his view resulted in decreased disease incidence and decreased fertilizer application with profuse tillering. The yield rose up from 5 tonnes to 6.5 tonnes in his field

Increased Area under Crops (Average)

The data related to increased area under crops has been collected and presented below in the Table 16.7.

Table 16.7: Increased Area under Crops (Average)

Name of the Crop	*Increased Area (ha) in IAMWARM Tanks*	*Percentage to Total Sample*	*Increased Area (ha) in Control Tanks*	*Percentage to Total Sample*
Paddy	0.45	62	0.14	26
Cotton	0.42	23	-0.12	-31
Chillies	0.22	46	-	-
Maize	0.68	64	0.11	31
Oilseeds	0.13	28	-	-
Pulses	0.16	34	0.10	12
Vegetable	0.44	51	0.21	37
Banana	0.50	53	0.13	28
Sugarcane	0.12	21	0.11	04

The IAMWARM Sub basins in Tamil Nadu have got farmers cultivating paddy, cotton, chillies, maize, oil seeds, pulses, banana, sugarcane and Vegetables. From the above table Maize is cultivated by most of the farmers and also the crop replaced the cotton area. It is found that the crops namely paddy, maize, vegetable and banana saw a more than 50 per cent increase in the area of cultivation after introduction of IAMWARM with the increased area under paddy and maize surpassed the other crops. Obviously, the impact of SRI in rice cultivation, Precision farming in Vegetables and banana also the resultant increase in income enabled the farmers to increase the area under other crops.

It is conspicuously seen from the above table that in control tanks where IAMWARM was not introduced, the increase in area under cultivation in major crops was not noticed and below except paddy and cotton which was also less pronounced when compared to treated tanks under IAMWARM

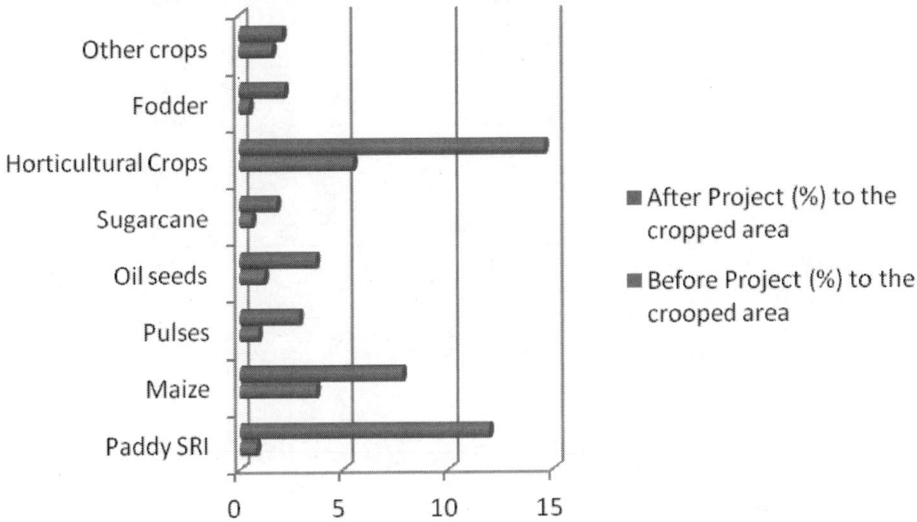

Cropping Pattern of the Project Tanks.

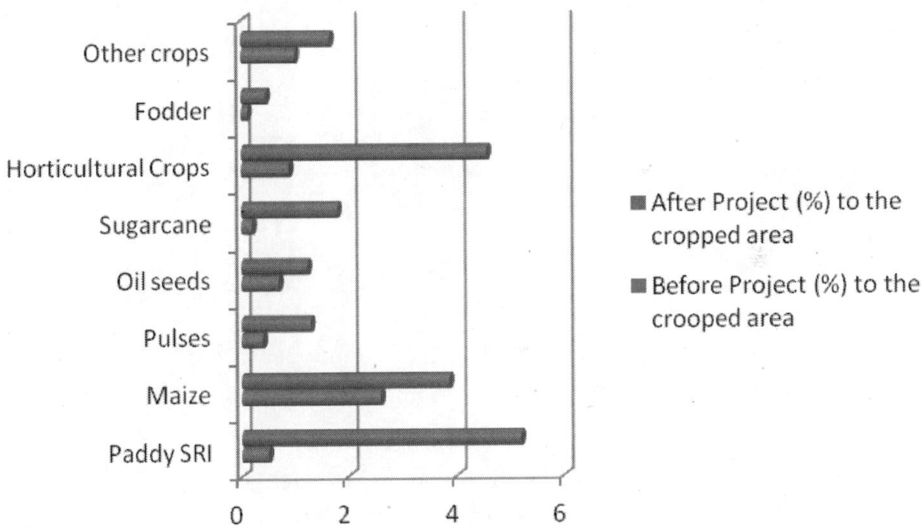

Cropping Pattern of the Control Tanks

The sustenance of Interventions was assessed and presented in the Table 16.10.

Among the interventions of TNAU and Department of Agriculture, the SRI in rice has been fairly successful and it was partially sustained even after the completion of the project. The obvious reason for this is the visible impact in terms of yield, reduced pest and disease incidence and profuse tillering.

Table 16.10: Sustenance of Interventions of the Study Area

Crop	*Name of the Interventions*	*Percentage to the Total Samples*
Paddy	SRI	57.27
Pulses	Rice Fallow pulses	2.68
Pulses	IPT	1.28
Maize	Hybrid seed	6.24
Vegetables	Precision farming	26.89
Banana	TC Banana	1.45
Groundnut	Sprinkler	1.63
Sugarcane	SSI	2.56

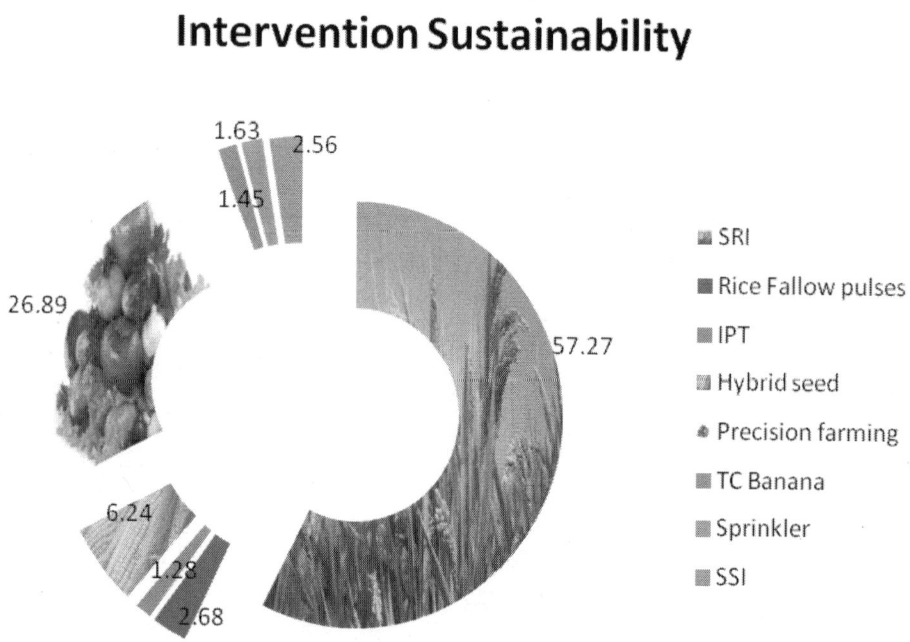

The precision farming intervention was very much successful during the implementation of IAMWARM and once the project was over, it was not sustained by the farmers for various reasons. The chief reason being the non- availability of water and frequent droughts.

Conclusion

The present paper proposed an approach to the evaluation of farm sustainability. The issue of the diversification and completely linked with it, the issue of gap area coverage in the Ayacut area also to be discussed. Depends upon the length of availability of water to the tail end and the total area that could be covered, the crop/ varieties are suggested for gap area sowing/coverage. When the water availability was shorter period, irrigated pulses of 70 -75 days duration suggested for the gap area. The water availability was for about 3 months, irrigated groundnut or irrigated millets were recommended for the gap area. When the water availability is more than 90 days in the gap area, maize, millets, *etc.*, were recommended based on WUA choice. These crops could supplement by well irrigation if available, depending on necessity. In the case of non availability of water for the tail end in the sub basin, rainfed millets; groundnut could be recommended in anticipating rain, instead of keeping the area fallow. The above said varieties and crops were sustained in the project area of TN-IAMWARM project sub basins.

As per the availability of water to the gap area, desired seeds and other inputs mobilized and stocked in the agricultural engineering for timely distribution to the farmers. The existing gap area in which wells are there, the medium duration irrigated Groundnut recommended with confidence, so that last one or two irrigation given through wells.

The TN-IAMWARM sub basin were farmers educated through publicity, village meeting, village campaigns, WUA meetings *etc.* about the climate resilient agriculture irrigated crops to taken up in the gap area depends upon the water availability or rainfed crops to taken up in the case of non-availability of water. Farmers were educated and followed that the gap area of the Ayacut should not left as fallow without any crop. Some crops whether fully irrigated or partially irrigated or partially substituted with well water or/pure rainfed crops including fodder crops could take up by the project beneficiaries. 10 to 20 per cent shift from paddy to other commercial crops, 30 to 40 per cent adopted of INM and IPM practices. 10 to 20 per cent increase in productivity. 10 to 20 per cent increase in farm income.

The extent of increase in the Horticulture crops found in the project area. The area shifted and sustained to diversified crop. The extent of adoption of INM and IPM practices. The area in which the soil fertility has improved. The extent increased in Bio- fertilizer application. The increase in productivity of the crops and the consequent additional agriculture income in the Sub-Basin. The notable change in the Lifestyle of the farmers due to the developmental activities in the Sub-Basin. The survey has shown that approaches of farm ranking could provide information at the educational level. Nevertheless, the research highlighted their usefulness also at the institutional level.

REFERENCES

John, M.A. and Flores, R. (2005) Impact evaluation of a conditional cash transfer program: The Nicaraguan Red de Social. Research Report 141, International Food Pol-icy Research Institute, Washington DC.

Namara, R.E., Upadhyay, B. and Nagar, R.K. (2005) Adoption and impacts of microirrigation technologies: Empirical results from selected localities of Maharashtra and Gujarat States of India. Research Report 93, International Water Management Institute, Colombo, Sri Lanka.

Narayanamoorthy, A. (2005) Economics of drip irrigation in sugarcane cultivation: Case study of a farmer from Tamil Nadu. *Indian Journal of Agricultural Economics*, **60**, 235-248.

Palanisami, K., Raman, S. and Mohan, K. (2012) Micro- irrigation Economics and Outreach. MacMillan Publishers India, New Delhi.

Palanisamy, K., "Sustainable Management on Tank Irrigation System in India", Bulletin 24, Water Technology Centre, Tamil Nadu Agricultural University Press, Coimbatore, 2002.

Pattanayak, K.S. (2009) Rough guide to impact evalua-tion of environmental and development program pro-grams. Sandee Working Paper No.40-09, South Asian Network for Development and Environmental Economics, Nepal.

Saleth, R.M. (1996) Water institutions in India: Econom-ics, law and policy. Commonwealth Publishers, New Delhi.

Seckler, D., Molden, U.A.D., de Silva, R. and Barker, R. (1998) World water demand and supply, 1990 to 2025: Scenarios and Issues. Research Report 19, International Water Management Institute (IWMI), Colombo, Sri Lanka.

Suresh Kumar, D. (2012) An analysis of economics of adoption of drip irrigation: Some experiences and evi-dences. The Bangladesh Development Studies, (In Press).

Vaidyanathan, A. (1999) Water resources management: Institutions and irrigation development in India. Oxford University Press, New Delhi.

Chapter 17

Impact of Climate Change on Insect Pests and Diseases

Saurav Gupta, Vinod Gupta, Neerja Sharma and A.K. Sinha*

KVK Samba, SKUAST-Jammu, J&K

The word insect comes from the Latin word *insectum*, meaning "cut up" or "divided into segments". The plural of *insectum*, namely *insecta*, is used by scientists as the name of the taxonomic class that insects belong to. This Latin word was created in order to translate the Greek word for "insect", which is *entomon*. This Greek word also literally means "cut up" or "divided into segments", and it is the source of the word *entomology*. The Greeks coined this term for insects because of the division of insect bodies into three segments, now called the head, thorax, and abdomen. Any animal of the class Insecta, comprising small, air-breathing arthropods having the body divided into three parts (head, thorax, and abdomen), and having three pairs of legs and usually two pairs of wings. Ants, beetles, butterflies, and flies are all insects. Insects are cold-blooded organisms - the temperature of their bodies is approximately the same as that of the environment. Therefore, temperature is probably the single most important environmental factor influencing insect behavior, distribution, development, survival, and reproduction. Insect life stage predictions are most often calculated using accumulated degree days from a base temperature and biofix point. Some researchers believe that the effect of temperature on insects largely overwhelms the effects of other environmental factors (Bale *et al.*, 2002). It has been estimated that with a 2° C temperature increase insects might experience one to five additional life cycles per season (Yamamura and Kiritani, 1998). Other researchers have found that moisture and CO_2 effects on insects can be potentially important considerations in a global climate change setting (Hamilton 2005, Coviella and Trumble 1999; Hunter, 2001).

The term DISEASE is coined by combining the words DIS + EASE = DISEASE. The prefix DIS means negative, reverse, or opposite, and the word EASE means comfort, or freedom from pain or discomfort. DIS-EASE therefore means not well, and the cause can be many. A plant disease may therefore be defined as: Any harmful deviation or alteration from the normal functioning of physiological processes. It is also defined by some as: Disease is a malfunctioning process that is caused by continuous irritation which results in suffering. A more practical definition of a disease would be: A plant is diseased when its systems are not normal and, therefore, it is not producing as well as it should according to normal expectations of the farmers. The pathogens, especially the animate ones, will not always be able to cause a disease unless environmental conditions and suitability of the host are also favourable for survival, multiplication, and entry of the pathogen into the plant and further development of the disease. For a pathogen to cause a disease, the pathogen has to be virulent in the first place; the host has to be susceptible, and the environmental conditions favourable. Disease will not develop if anyone of the 3 conditions is not fulfilled. This is known as the disease triangle. The mere presence of pathogen and its susceptible host in the field does not necessarily mean that disease will develop. The environmental factors which most commonly affect the initiation and development of infectious plant diseases are temperature and moisture on the plant surface Soil Nutrients also play an important role in some diseases, and to lesser extent, light, soil pH and oxygen and carbon dioxide levels, particularly in soil environment also affect disease development. These factors affect disease development through their influence on the growth and susceptibility of the host, on the multiplication and activity of the pathogen as it is relevant to the severity of symptom development. For a disease to occur and to develop optimally, a combination of three factors must be present susceptible plant, infective (virulent) pathogen, and favourable environment. Although plant susceptibility and pathogen virulence remain essentially unchanged in the same plant for at least several days, weeks or months, the environmental conditions may change more or less drastically. The major predicted results of climate change increases in temperature, moisture and CO2 – can impact all three legs of the plant disease triangle in various ways. Precisely predicting the impact of climate change on plant disease is tricky business.

Effect of Rising Temperature on Plant Diseases and Insect Pests

Temperature has potential impacts on plant disease through both the host crop plant and the pathogen. Research has shown that host plants such as wheat and oats become more susceptible to rust diseases with increased temperature; but some forage species become more resistant to fungi with increased temperature (Coakley *et al.,* 1999).Seasonal and regional occurrence of disease is to a great extent, determined by temperature. The too low temperatures of winter and very high temperatures of summer are, below the minimum and above the maximum respectively required by most pathogens. Therefore, diseases may not be initiated and those in progress may come to a halt during these periods of the year in our country. With the advent of favourable temperature pathogens become active and if other conditions are favourable they can cause the disease. Pathogens differ for their choice of low or high temperatures. Thus some species of *Typhula* (snow molds of cereals and turf grasses) thrive only in cool seasons or cold areas. Peach leaf curl caused by *Taphrina deformans* occurs only in cool (temperate) areas. This plant is grown in India both in hills as well as the plains. But the disease is found only in the hills, assuming severe state during cold and humid spring. Yellow rust of wheat and stripe disease of barley are also favoured by low temperature. The late blight pathogen, *Phytopthora infestans* is most destructive only during winter in our country and other sub tropical areas. Some diseases on the other hand are favoured by high temperatures and thus are common in the warm climate of tropics and subtropics. Thus *Fusarial wilts*, the *Phymatotrichum* root rots, the brown rot of stone solanaceous plants (*Pseudomonas solanacearum*) and *anthracnoses* caused by *colletotrichum* are more prevalent in warmer areas. *Pseudomonas solancearum*, causing bacterial wilt and brown rot of potato does not affect the potato crop in the plains of India which is raised during winter but may occur when the crop is planted during February. Even on the hills of north India the disease is more common in the valleys (warmer) than on higher altitudes. The soft rot of potato (*Erwinia* spp.) is also common in warm, humid seasons. *Sclerotium rolfsii*, causing rot in many crop plants is destructive in warmer areas.

In blast disease of rice (*Pyricularia oryzae*) plant age and host nutrition become of secondary importance when temperature and moisture are favourable for stages in the disease development *viz.* production and germination of conidia, dispersal, host resistance alteration and nutrition. As a group, the powdery mildews have an optimum temperature for conidia germination and growth of about 21°C. The late blight for potato is reported to be unknown in areas where the mean atmospheric temperature exceeds 25°C. The temperature affects disease incidence and its spread also indirectly through its effect on insect vectors. Temperature, in combination with sunlight may determine seasonal appearance of symptoms in various viral diseases. Viruses causing yellows or leaf- rolls are most severe in summer whereas those causing mosaics or ring spots during spring or winter.

Climate change resulting in increased temperature could impact crop pest insect populations in several complex ways. Although some climate change temperature effects might tend to depress insect populations, most researchers seem to agree that warmer temperatures in temperate climates will result in more types and higher populations of insects. Researchers have shown that increased temperatures can potentially affect insect survival, development, geographic range, and population size. Temperature can impact insect physiology and development directly or indirectly through the physiology or existence of hosts. Depending on the development "strategy" of an insect species, temperature can exert different effects (Bale *et al.*, 2002). Some insects take several years to complete one life cycle – these insects (cicadas, arctic moths) will tend to moderate temperature variability over the course of their life history. Some crop pests are "stop and go" developers in relation to temperature – they develop more rapidly during periods of time with suitable temperatures. Use of degree-day or phenology based models to predict the emergence of these insects and their potential to damage crops (cabbage maggot, onion maggot, European corn borer, Colorado potato beetle). Increased temperatures will accelerate the development of these types of insects – possibly resulting in more generations (and crop damage) per year. "Migratory" insects (corn earworm in northern parts of the northeast) may arrive in the Northeast earlier, or the area in which they are able to overwinter may be expanded. Natural enemy and host insect populations may respond differently to changes in temperature. Parasitism could be reduced if host populations emerge and pass through vulnerable life stages before parasitoids emerge.

Hosts may pass though vulnerable life stages more quickly at higher temperatures, reducing the window of opportunity for parasitism. Temperature may change gender ratios of some pest species such as thrips (Lewis 1997) potentially affecting reproduction rates. Insects that spend important parts of their life histories in the soil may be more gradually affected by temperature changes than those that are above ground simply because soil provides an insulating medium that will tend to buffer temperature changes more than the air (Bale *et al.*, 2002). Lower winter mortality of insects due to warmer winter temperatures could be important in increasing insect populations (Harrington *et al.*, 2001). Higher average temperature might result in some crops being able to be grown in some regions – it is likely that at least some of the insect pests of those crops will follow the expanded crop areas. Insect species diversity per area tends to decrease with higher latitude and altitude (Gaston and Williams 1996, Andrew and Hughes 2005), meaning that rising temperatures could result in more insect species attacking more hosts in temperate climates (Bale *et al.*, 2002). Based on evidence developed by studying the fossil record some researchers (Bale *et al.*, 2002) conclude that the diversity of insect species and the intensity of their feeding have increased historically with increasing temperature.

Some insects are closely tied to a specific set of host crops. Temperature increases that cause farmers not to grow the host crop any longer would decrease the populations of insect pests specific to those crops. The same environmental factors that impact pest insects can impact their insect predators and parasites as well as the disease organisms that infect the pests,resulting in increased attack on insect populations. At higher temperatures, aphids have been shown to be less responsive to the aphid alarm pheromone they release when under attack by insect predators and parasitoids – resulting in the potential for greater predation (Awmack, 1997).

Effect of Moisture on Disease Development and Insect Pest's Incidence in Crops

Moisture also effects the initiation and development of diseases in many interrelated ways. The most important influence is on the germination of fungal spores and penetration of host by germ tube. Moisture also activates the bacterial, fungal and nematodal pathogens. As splashing rain and running water, moisture also plays important role in distribution and spread of many pathogens, on the same plant or from one plant to another. Moisture also increases the succulence of host tissues.

The occurrence of many diseases only in particular region is closely correlated with the amount and distribution of rainfall within the year. Thus, late blight of potato, apple scab, downy mildews, fire blight *etc.* rarely occur in dry areas and found or are severe only in areas with high rainfall or high relative humidity during the growing season. Most fungal pathogens depend on the presence of free moisture on the host or of high relative humidity in atmosphere only during the spore germination, and become independent once they can obtain nutrients and water from the host. Some, as late blight of potato and downy mildews, however, require high relative humidity in the air throughout their development. These diseases come to a halt as soon as dry, hot weather sets in n resume only after a rain or after the return of humid weather.

Contrary to most fungal and bacterial pathogens of aboveground plant parts (that require water film), the spores of powdery mildews can germinate, penetrate and cause infection when there is high relative humidity in the atmosphere surrounding the plant. In some powdery mildews, most severe infections occur when relative humidity is rather low (50-70 per cent). Thus powdery mildews are more common and more severe in drier areas of the world. Most bacterial diseases and many fungal diseases of young tender tissues are favoured by high moisture or high relative humidity.

Some example of the diseases favoured by dry soil or water deficiency are, seedling blight of cereals (*Fusarium roseum*), common scab of potato (*Streptomyces scabies*) root and stem rot of peas (*F. solani* f. sppisi), charcoal rot of jowar, potato, cotton *etc.* (*Macrop Hominap Hoseolina*), stalk rot of maize (*Fusarium moniliforme*), rice blast (*Pyricularia oryzae*), *etc.* Some example of the diseases favoured by wet soil or high relative humidity are root rot of cotton (*PHymatotrichum omnivorum*), take all of wheat (*Gaeumannomyces graminis* var. *tritici*), stripe of barley (*Drechsler agraminea*), root rots (*Sclerotium rolfsii*), black root rot of tobacco (*Thielaviopsis basicola*), sclerotinia diseases of vegetables and field crops (*Sclerotinia sclerotiorum*), wart of potato (*Synchytrium endobioticum*), damping off of seedlings (*Pythium* spp.) late blight of potato (*Phytophthora infestans*), root-knot of many plants (*Meloidogyne incognita*), bacterial wilt of tomato and potato (*Pseudomonas solanacearum*) *etc.*

There are fewer scientific studies on the effect of precipitation on insects than temperature. Some insects are sensitive to precipitation and are killed or removed from crops by heavy rains, this consideration is important when choosing management options for onion thrips (Reiners and Petzoldt, 2005). For some insects that overwinter in soil, such as the cranberry fruit worm and other cranberry insect pests, flooding the soil has been used as a control measure (Vincent *et al.*, 2003). One would expect the predicted more frequent and intense precipitation events forecasted with climate change to negatively impact these insects. Other insects such as pea aphids are not tolerant of drought (Macvean and Dixon 2001). As with temperature, precipitation changes can impact insect pest predators, parasites, and diseases resulting in a complex dynamic. Fungal pathogens of insects are favored by high humidity and their incidence would be increased by climate changes that lengthen periods of high humidity and reduced by those that result in drier conditions.

Effect of CO_2 Levels on Disease and Insect Pests in Crops

Increased CO_2 levels can impact both the host and the pathogen in multiple ways. Some of the observed CO_2 effects on disease may counteract others.

Researchers have shown that higher growth rates of leaves and stems observed for plants grown under high CO_2 concentrations may result in denser canopies with higher humidity that favor pathogens. Lower plant decomposition rates observed in high CO_2 situations could increase the crop residue on which disease organisms can overwinter, resulting in higher inoculums levels at the beginning of the growing season, and earlier and faster disease epidemics. Pathogen growth can be affected by higher CO_2 concentrations resulting in greater fungal spore production. However, increased CO_2 can result in physiological changes to the host plant that can increase host resistance to pathogens (Coakley *et al.*, 1999).Host resistance to disease may be overcome more quickly by more rapid disease cycles, resulting in a greater chance of pathogens evolving to overcome host plant resistance. Fungicide and bactericide efficacy may change with increased CO_2, moisture, and temperature. The more frequent rainfall events predicted by climate change models could result it difficult to keep residues of contact fungicides on plants, triggering more frequent applications. Systemic fungicides could be affected negatively by physiological changes that slow uptake rates, such as smaller stomata opening or thicker epicuticular waxes in crop plants grown under higher temperatures. These same fungicides could be affected positively by increased plant metabolic rates that could increase fungicide uptake. Exclusion of pathogens and quarantines through regulatory means may become more difficult for authorities as unexpected pathogens might appear more frequently on imported crops.

Generally CO_2 impacts on insects are thought to be indirect - impact on insect damage results from changes in the host crop. Some researchers have found that rising CO_2 can potentially have important effects on insect pest problems. Recently, free air gas concentration enrichment (FACE) technology was used to create an atmosphere with CO_2 and O_2 concentrations similar to what climate change models predict for the middle of the 21st century. FACE allows for field testing of crop situations with fewer limitations than those conducted in enclosed spaces. During the early season, soybeans grown in elevated CO_2 atmosphere had 57 per cent more damage from insects (primarily Japanese beetle, potato leafhopper, western corn rootworm and Mexican bean beetle) than those grown in today's atmosphere, and required an insecticide treatment in order to continue the experiment. It is thought that measured increases in the levels of simple sugars in the soybean leaves may have stimulated the additional insect feeding (Hamilton *et al.*, 2005). It has been also observed that insects sometimes feed more on leaves that have lowered nitrogen content in order to obtain sufficient nitrogen for their metabolism (Coviella and Trumble 1999, Hunter 2001). Increased carbon to nitrogen ratios in plant tissue resulting from increased CO_2 levels may slow insect development and increase the length of life stages vulnerable to attack by parasitoids (Coviella and Trumble. 1999). Entomologists expect that insects will expand their geographic ranges, and increase reproduction rates and overwintering success. It is apparent that for sweet corn pests, warmer temperatures translate to increased insecticide applications to produce a marketable crop. Insecticides and their applications have significant economic costs for growers and environmental costs for society. Additionally, some classes of pesticides (pyrethroids and spinosad) have been shown to be less effective in controlling insects at higher temperatures (Musser and Shelton 2005).

Entomologists predict additional generations of important pest insects in temperate climates as a result of increased temperatures, probably necessitating more insecticide applications to maintain populations below economic damage thresholds. A basic rule of thumb for avoiding the development of insecticide resistance is to apply insecticides with a particular mode of action less frequently (Shelton *et al.*, 2001, Georghiou and Taylor 1986). With more insecticide applications required, the probability of applying a given mode of action insecticide more times in a season will increase, thus increasing the probability of insects developing resistance to insecticides. In New York a network of pheromone traps in sweet corn fields has been used to monitor corn earworm (*Helicoverpa zea*) throughout the central and western part of the state for over 10 years (Seaman personal communication). Corn earworm is thought not to overwinter in upstate New York and is generally considered to be a late season, migratory pest of sweet corn, so trapping was initiated in mid-July. A number of cultural practices that can be used by farmers could be affected by changes in climate. Using crop rotation as an insect management strategy could be less effective with earlier insect arrival or increased overwintering of insects. However, this could be balanced by changes in the earliness of crop planting times, development, and harvest.

Management Strategies

Disease management strategies depend on climate conditions. Climate change will cause alterations in the disease geographical and temporal distributions and consequently the control methods will have to be adapted to this new reality. There are few discussions on how chemical control will be affected by climate change, despite the importance of this subject. Changes in temperature and precipitation can alter fungicide residue dynamics in the foliage, and the degradation of products can be modified. Alterations in plant morphology or physiology, resulting from growth in a CO_2-enriched atmosphere or from different temperature and precipitation conditions, can affect the penetration, translocation and mode of action of systemic fungicides. Besides, those changes in plant growth can alter the period of higher susceptibility to pathogens which can determine a new fungicide application calendar (Coakley *et al.*, 1999; Chakraborty and Pangga, 2004). The main impact of climate change on chemical control will be in the cultural realm. The fact that the entire humanity is suffering the consequences of anthropogenic activity in the process of exploiting the resources of the planet will raise awareness that this activity must be conducted in a sustainable way. Society will certainly exert pressure for the use of non-chemical methods to control plant diseases. One of the direct consequences of climate change in the pathogen-host relationship is the genetic resistance of plants to diseases. Many changes in plant physiology can alter the resistance mechanisms of cultivars obtained by both traditional and genetic engineering methods. Current agricultural practices may need to be revisited because of climate change. IPM is a widely used strategy that integrates cultural, biological, and chemical controls to reduce harmful insect populations below a threshold which otherwise will cause economic losses. In addition, as temperatures increase, the frequency of spring frosts will decline and the resulting extended frost free periods will increase the duration and intensity of insect outbreaks. Indeed, farmers may be tempted to take advantage of the changing climate by planting crops earlier in the season. However, plants will then be available for insect pests earlier in the annual cycle, allowing populations to get a quicker start and potentially add additional generations during a typical growing season (Trumble and Butler 2009). It could translate to increased insect and pathogen pressure at the end of the season, resulting in higher associated yield losses. In some cases, small differences in temperature and humidity can determine the preferred protection strategy. IPM less reliant on chemical crop protection can mutatis mutandis be raised to develop locally adaptive and diversified IPM systems which are resilient enough to meet extreme weather fluctuations as a result of climate change. Farmers should keep in mind that climate change is likely to be a gradual process that will give them some opportunity to adapt. Farmers who closely monitor the occurrence of pests in their fields and keep records of the severity, frequency, and cost of managing pests over time will be in a better position to make decisions about whether it remains economical to continue to grow a particular crop or use a certain pest management technique. If more fungicide or insecticide applications are required in order to successfully grow a particular crop, farmers will need to carefully evaluate whether growing that crop remains economical. Those farmers who make the best use of the basics of integrated pest management (IPM) such as field monitoring, pest forecasting, record keeping, and choosing economically and environmentally sound control measures will be most likely to be successful in dealing with the effects of climate change

Summary

The precise impacts of climate change on insects and pathogens are somewhat uncertain because some climate changes may favour pathogens and insects while others may inhibit a few insects and pathogens. The preponderance of evidence indicates that there will be an overall increase in the number of outbreaks of a wider variety of insects and pathogens. The possible increased use of fungicides and insecticides resulting from an increase in pest outbreaks will likely have negative environmental and economic impacts for agriculture in the world. The best economic strategy to overcome the effects of climate change is to follow the use of integrated pest management practices to closely monitor insect and disease occurrence. Keeping pest and crop management records over time will allow farmers to evaluate the economics and environmental impact of pest control and determine the feasibility of using certain pest management strategies or growing particular crops.

REFERENCES

Andrew, N.R. and L. Hughes. 2005. Diversity and assemblage structure of phytophagous Hemiptera along a latitudinal gradient: predicting the potential impacts of climate change. *Global Ecol Biogeogr*. 14: 249-262.

Awmack, C.S., C.M. Woodcock and R. Harrington. 1997. Climate change may increase vulnerability of aphids to natural enemies. Ecological Entomology. 22: 366-368.

Bale, J.S. G.J. Masters, I.D. Hodkinson, C. Awmack, T.M. Bezemer, V.K. Brown, J. Butterfield, A. Buse, J.C. Coulson, J. Farrar, J.*E.G.* Good, R. Harrington, S. Hartley, T.H. Jones. R.L. Lindroth, M.C. Press, I. Symrnioudis, A.D. Watt, and J.B. Whittaker. 2002. Herbivory in global climate change research: direct effects of rising temperatures on insect herbivores. *Global Change Biology* 8: 1-16.

Chakraborty, S., A.V. Tiedemann, P.S. Teng. 2000. Climate Change: Potential impact on plant diseases. Environ. Poll. 108: 317-326.

Coakley,S.M., H. Scherm, S. Chakraborty. 1999. Climate Change and Disease Management. *Ann. Rev. Phyto*. 37: 399-426.

Coviella, C. and J. Trumble. 1999. Effects of elevated atmospheric carbon dioxide on insect plant interactions. *Conserv. Biol*. 13: 700-712.

Gaston,K.J. and P.H. Williams. 1996. Spatial patterns in taxonomic diversity. In: *Biodiversity* 202-229. Blackwell Science, Oxford.

Hamilton, J.G., O. Dermody, M. Aldea, A.R. Zangerl, A. Rogers, M.R. Berenbaum, and E.Delucia. 2005. Anthropogenic Changes in Tropospheric Composition Increase Susceptibility of Soybean to Insect Herbivory. *Envirn.Entomol*. 34: 2 479-485.

Harrington, R., R, Fleming, I. P. Woiwood. 2001. Climate change impacts on insect management and conservation in temperate regions: can they be predicted? *Agricultural and Forest Entomology*. 3: 233-240.

Hunter, M.D. 2001. Effects of elevated atmospheric carbon dioxide on insect-plant interactions.*Ag.Forest.Entomo*l. 3: 153-159.

Lewis, T. 1997. Thrips as crop pests.*CAB International, Cambridge: University Press*. 740pp.

Mcvean, R. and A. F. G. Dixon. 2001. The effect of plant drought-stress on populations of the pea aphid *Acyrthosi phonpisum*. Ecol. Entomol. 26: 440-443.

Musser, F. P and A. M. Shelton. 2005. The influence of post-exposure temperature on the toxicity of insecticides to *Ostrinia nubilalis* (Lepidoptera: Crambidae). Pest Manag Sci. 61: 508- 510.Georghiou, G.P. and Taylor, C.E. 1986. Factors influencing the evolution of resistance.In: Pesticide Resistance: Strategies and tactics for management. National Research Council, National Academy Press. Pages 143-157.

Reiners, S and C. Petzoldt (eds). 2005. Integrated Crop and Pest Management Guidelines for Commercial Vegetable Production. Cornell Cooperative Extension publication

Shelton, A.M., W.R. Wilsey, and D.M. Soderlund. 2001. Classification of insecticides and acaricides for resistance management. Dept. of Entomology, NYSAES, Geneva, NY 14456.

Trumble JT, Butler CD (2009) Climate change will exacerbate California's insect pest problems. Calif Agric 63: 73–78

Vincent, C., G. Hallman, B. Panneton, and F. Fleurat-Lessardú. 2003. Management of agricultural insects with physical control methods. Ann Rev Entomol 48: 261-281.

Yamamura, K. and K. Kiritani. 1998. A simple method to estimate the potential increase in the number of generations under global warming in temperate zones. *Appl. Ent. and Zool.* 33: 289- 298.

Chapter 18

Radar Based Rainfall Measurement: A Brief Review

Md. Majeed Pasha and K.K. Sathian*

Department of Soil and Water Conservation Engineering, Kelappaji College of Agricultural Engineering and Technology, Tavanur, Kerala

Introduction

Precipitation can be measured and represented in many ways such as point data from rain gauges, grid data from meteorological radar, or interpolated data. Precipitation is a key control on watershed hydrologic modelling output, with errors in rainfall propagating through subsequent stages of water quantity and quality analysis. Hydrological analysis can be performed with confidence only if the precipitation input is adequately measured in spatial and temporal extents. An extremely dense rain gauge network is needed for efficient rainfall-runoff estimation of a watershed. The dense network of 2 × 2 km might be required for planning purposes in India Suresh (2010). However, owing to the practical difficulties, an acceptable density of the rain gauge network was proposed as 4 × 4 km but this network density is also functionally impractical in view of topographical and terrain constraints, cost restrictions and maintenance related problems. At present, India has one rain gauge for every 3402 sq km on an average (Guhathakurta and Rajeevan, 2008).

The accuracy of hydrologic model simulations depends mainly on the accuracy of model inputs such as precipitation. Most watershed models incorporate precipitation data from rain gauges; Even though gauged rainfall data generally provide accurate depth measurements, sparsely spaced, gauging stations cannot effectively account for the spatial variability of precipitation at basin scale. Higher resolution data sources are available, but they are associated with greater computational requirements and expertise. Watershed hydrologic models are widely used to simulate. It is well-known that rainfall is a highly heterogeneous process covering an extensive range of scales in time and space (Marani, 2005; Nicótina, 2008). Despite the fact that accurate estimates of rainfall in terms of location and intensity are crucial for operational water management, as well as for the hydrological research community (Van de Beek *et al.*, 2010), hydrologists have traditionally put more effort in the development of evermore sophisticated rainfall-runoff modeling approaches than in the development of improved techniques for the measurement and prediction of the space–time variability of rainfall, streamflow for water quality and ecosystem services and to evaluate alternative future scenarios associated with land use and climate change. However, capturing precipitation spatial and temporal variability successfully from available resources has been a challenge for hydrologic modelling.

The main sources of precipitation data are rain gauges, Doppler weather radar, and satellite sensors, with many subcategories of each, and each data source has associated specific considerations of instrumental uncertainty and spatio-temporal resolution (Moreau *et al.*, 2009). For practical applications of water resource management and flood, forecasting, lumped models are commonly employed. Current lumped models use averages for meteorological inputs and watershed characteristics and such are prone to calibration and extrapolation difficulties in the ungauged or poorly gauged basins. The spatial pattern of precipitation input cannot be translated into the simulated hydrograph by the lumped models. However, distributed models incorporate the temporal and spatial variability of watershed and meteorological inputs and allow better representations of the hydrological processes compared with other model types. Developments of the distributed hydrological models have emphasized capturing the local unique contributions due to different land cover, soil property and topography, and the sensitivity analysis of surface spatial heterogeneity (Yang *et al.*, 2001). Spatial variability of precipitation impacts the predictions of the hydrological processes such as seasonal flow in streams, floods, evapotranspiration, recharge, and ground water heads. Moreover, it is of significant importance for the closure of the water balance of a watershed or a region (Vischel and Lebel, 2007).

Methods of Measuring Rainfall

Raingauge

The purpose of the raingauge is to measure the depth and intensity of rainfalling on a flat surface without considering infiltration, runoff or evaporation. The problems of measurements include effects of topography, nearby vegetation and the design of gauge itself.

Types of Raingauge

There are mainly two types of raingauge (non-recording and recording).

1. Non-recording Gauge
2. Symons raingauge

Recording Type Raingauge

- Weighting Bucket Type
- Raingauge, Tipping Bucket Type
- Raingauge and Siphon Type
- Automatic Rainfall Recorder

Radars

Radar stands for Radio Detection and Ranging. It refers to the use of radio waves to detect objects and determine the distance (range) to the object. Modern weather radars are mostly pulse-Doppler radars, capable of detecting the motion of rain droplets in addition to the intensity of the precipitation. Both types of data can be analyzed to determine the structure of storms and their potential to cause severe weather. During World War II, radar operators discovered that weather was causing echoes on their screen, masking potential enemy targets. Techniques were developed to filter them, but scientists began to study the phenomenon. Soon after the war, surplus radars were used to detect precipitation. Since then, weather radar has evolved on its own and is now used by national weather services, research departments in universities, and in television stations' weather departments. Raw images are routinely used and specialized software can take radar data to make short term forecasts of future positions and intensities of rain, snow, hail, and other weather phenomena. Meteorological long range radars can be extremely useful for the observation of rainfall events over a wide area. Radar's ability to provide detailed information, in both time and space with precipitation patterns is unsurpassed by rain gauges.

Radar weather information are used to determine current atmospheric conditions, but meteorologists must integrate data from many sources to get a complete and accurate picture of atmospheric conditions. Meteorologists feed current data into computer models to help them predict weather conditions (forecast) and make critical decisions. Radar output is even incorporated into numerical weather prediction models to improve analyses and forecasts.

Importance of Radars

- They can see through fog, cloud, rain and other types of atmospheric conditions which light cannot pass through.
- They can observe many places in the sky almost simultaneously.
- They can run continuously, often without operators being present (computer controlled).
- They can operate during both day and night.
- The data from them can be easily stored to computer and then subjected to many types of sophisticated analysis.
- Modern solid-state radars often need little maintenance, or replacement of parts, so the largest expense is often the setup costs.
- Radars are not restricted to ground-level studies, and can often observe several kilometres upward into the atmosphere

Weather Radar in Hydrological Applications

- The weather radar data are available at fine spatial and temporal resolutions, and are suitable for areas where rain gauges are inadequate or absent.
- Radars assist the meteorologist to predict the forthcoming storms before they strike the catchment of interest.
- Forecasting the flash flood and severe thunderstorm could be possible using radar data.
- Existing rain gauges can be supplemented with the radar rainfall data.
- The study on rainfall variability on a micro scale is possible with radar data.
- Precipitation type and intensity could be determined.
- The watershed management could be planned in a better way based on the rainfall estimation over the catchments.

Basic Operation Principle

Electromagnetic energy is transmitted into the atmosphere. Once reaching a target (cloud droplets, ice crystals, raindrops, snowflakes, aerosol particles, insects, birds, airplane, *etc.*), the energy is absorbed and scattered. A portion of backscattered energy is received and processed by a radar to display as an echo

Important Radar Parameter

Peak Power

Instantaneous power emitted in a pulse) $10 < Pt < 5000$ kW, 5×10^6 W. Minimum detectable signal Much smaller than emitted energy ~ 10^{-13} W Because of the large range of the energy dealt with in a radar system (10^{-13} ~10^6), the power is often expressed in decibels (dB). Difference between two power level p_1 and p_0 is given by

$$p(dB) = 10 \log_{10}(p_1/p_0)$$

So, the dynamic range of a radar ~ 190 dB in electronic term, $p_o = 1$ mW

(10^{-3} W), unit dBm Minimum detectable ~ -100 dBm Peak power ~ 90 dBm

Weather Radar Classification Based on the Frequency Band

Different Bands	*Wavelength (cm)*	*Frequency (GHz)*	*Used for*
L	15-30	1-2	Clear air turbulence studies.
S	8-15	2-4	Near and far range weather observation since the S-band radars are not easily attenuated. The drawback of this band of radar is that it requires a large antenna dish and a large motor to power it.
C	4-8	4-8	Short range weather observation and TV stations as the dish size is small.

Different Bands	*Wavelength (cm)*	*Frequency (GHz)*	*Used for*
X	2.5-4	8-12	Studies on cloud development because they are more sensitive and can detect the tiny water particles and snow. Most major airplanes are equipped with X band radar to pick up turbulence and other weather phenomenon.
K	0.75-1.2or1.7-2.5	27-40or12-18	This band is similar to the X band, but it is more sensitive.

Radio Frequency (v)

Radio wavelength - λ - (v = c/λ) 3 < v<30 GHz (1 GHz = 109 $sec^{-1)}$ (wavelengths from 1 to 30 cm). Detection capability of hydrometeor depends critically on radio frequency/wavelength. In general, the smaller the size of the particles, the shorter the wavelength required to detect, *e.g.* the popular 10 cm (3 GHz or S-band) radar can detect rain drops but not cloud droplets which may be detected by 95 or 35 GHz radar.

Pulse Repetition Frequency (PRF)

Typical PRF for weather radar fr = 1000 s^{-1} but may range between 200 – 2000 s^{-1} Maximum range of detection for a radar set half the interval between pulses times the speed of light: c/2fr ~ 150 km for fr = 1000 s^{-1} range: 750km < c/2fr < 75 km

Pulse Duration

0.1 < τ< 5 µs vs. pulse interval 500 < t < 5000 µsec

Beam Width

Angular separation between points where the transmitted intensity has fallen to 1/2 its maximum value (*i.e.* 3 dB below the maximum). The smaller the beam width, the better the resolution. Typical width for weather radar 1°-3°

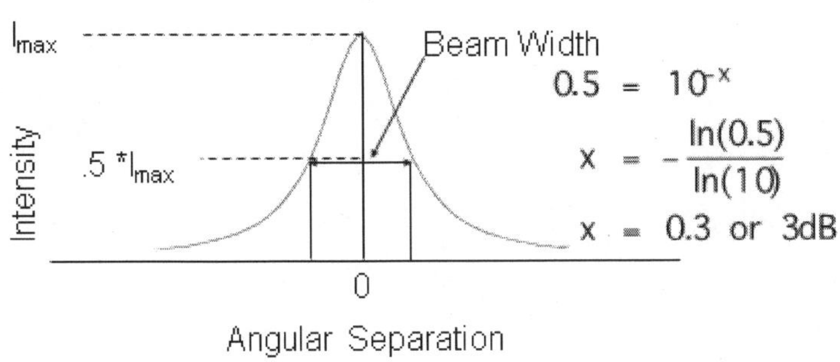

Signal Processing and Product Generation in Radars

The processing of radar data generally involve two distinct steps. The first step, called signal processing, is the extraction of raw radar parameters like echo strength (reflectivity) or Doppler velocity from the radar signals coming out of the receiver. The second step, called data processing or product generation, is the further processing of raw radar parameters in order to obtain information that is useful for meteorological or hydrological purposes. In general, these two steps are done by different computers, signal processing being done at the radar site, while product generation can be done everywhere the data are sent to.

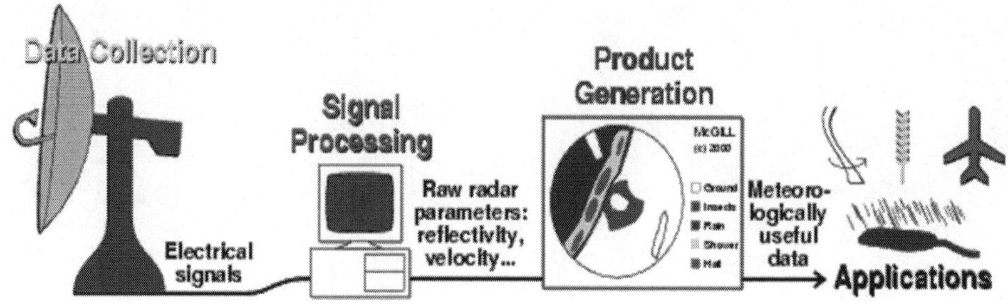

Basic Components of Radar System

Transmitter

Produces high power pulses at desired frequency. Pulses may be 1 microsecond in duration.

Receiver

Detects, amplifies and converts (digitizes) received voltages from each pulse as a function of range.

Antenna

Radiates transmitted power in narrow beam for maximum "gain receives backscattered signal from targets.

T/R Switch

Switches antenna between transmitter and receiver at high rate, typically once every millisecond, receivers are designed to detect very weak signals, on the order of 10-13 to 10-14 watts. Transmitted power is typically 106 watts, peak power. More advanced radars provide either frequency agility or polarization agility.

Frequency Agility

Transmit and receive multiple frequencies (near in frequency to each other) to increase sampling rate.

Polarization Agility

Alternatively transmit either horizontal or vertical polarization to provide wealth of information on particle phase, shape and orientation.

Sending and Receiving Signals and Detecting a Target

The radar creates an electromagnetic energy pulse which is focused by an antenna and transmitted through the atmosphere. Objects in the path of this electromagnetic pulse, called targets, scatter the electromagnetic energy. Some of that energy is scattered back toward the radar.

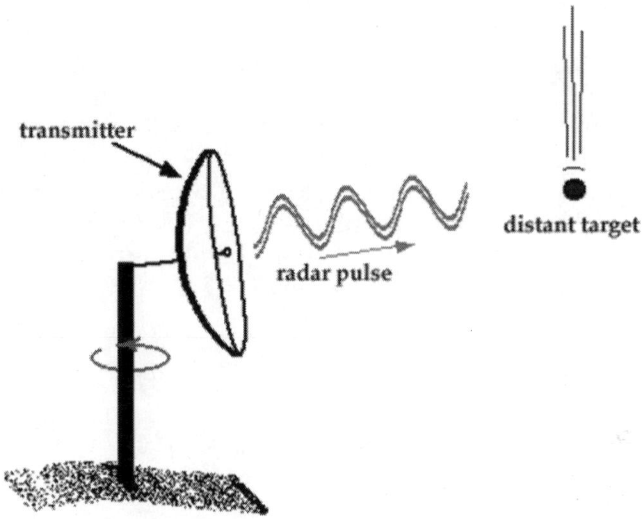

The receiving antenna (which is normally also the transmitting antenna) gathers this back-scattered radiation and feeds it to a device called a receiver.

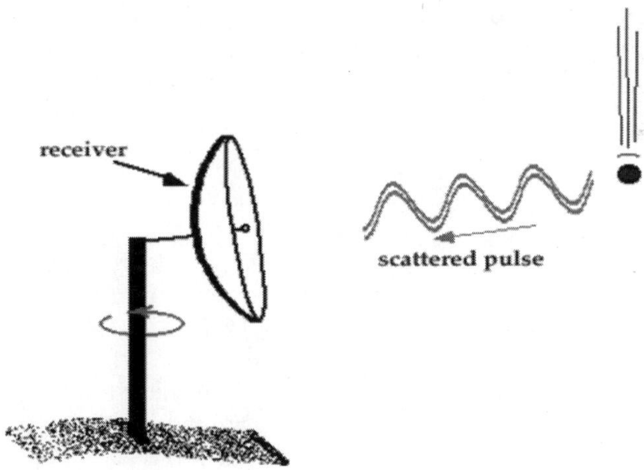

Different Types of Radar

Standard Weather Watch

Reflectivity only (Rain, drizzle, smoke, swaying trees on mountains, birds and bats).

Coherent or Doppler Radar

Doppler capability allows radar systems to provide information on the radial velocity of the backscatters. This is derived from the frequency shift induced by the Doppler effect. Operational weather radars such as the S-band WRS-88D in the USA and many C-band radars in Europe have this Doppler capability. There are two main Doppler variables commonly measured by Doppler radar systems. The mean Doppler radial velocity V [m s^{-1}] is the mean of the distribution of the velocity of the hydrometeors within the sampling volume along the radar beam direction. It is mainly influenced by the mean wind velocity, although hydrometeors are not passive tracers of the atmospheric wind field. The Doppler spectrum width rV [m s^{-1}] is the standard deviation of the distribution of the velocity of the hydrometeors within the sampling volume along the radar beam direction. It is sensitive to processes such as atmospheric turbulence that can significantly modify the velocity distribution.

Range and Doppler Measurements in RADAR Systems

$$E_t(t) = E_0(t_0)\cos\left[2\pi f_0 t + \varphi_0(t)\right]$$

E_0 = Amplitude of the signal

f_0 = RF frequency of the signal

φ_0 = Phase of the signal (possible modulated)

The returned signal is delayed by the time that takes to signal to reach the target and to return back to the receiver. Since the electromagnetic waves travel with the speed of light c (much greater then RADAR and Target velocities), the received signal is delayed by

$$t_d \cong \frac{R_1 + R_2}{c}$$

The received signal is:

$$E_r(t) = \alpha E_0(t - t_d)\cos\left[2\pi f_0(t - t_d) + \varphi(t - t_d)\right] + noise(t)$$

$\alpha < 1$ represents the attenuation of the signal To retrieve the range (and range-rate) information from the received signal the transmitted signal must be modulated in Amplitude or/and Frequency or/and Phase. The received signal is:

$$E_r(t) = \alpha E_0(t - t_d)\cos\left[2\pi f_0(t - t_d) + \varphi(t - t_d)\right] + noise(t)$$

We want to compute the delay time td due to the time td_1 it takes the EM-wave to reach the target at a distance R_1 (at t=0), from the transmitter, and to the time td_2 it takes the EM-wave to return to the receiver, at a distance R2 (at t=0) from the target. $T_d=td_1+td_2$

$$R_1(t) = R_1 + R_1.t \; R_2(t) = R_2 + R_2.t$$

According to the Special Relativity Theory the EM wave will travel with a constant velocity c (independent of the relative velocities).

$$R_1 \text{ and } R_2$$

Polarimetric Radar

Falling hydrometeors are generally not spherical. In particular, raindrops with an equivolume diameter larger than about 1 mm are oblate: their horizontal dimension is larger than their vertical one. This oblateness implies that a radar signal linearly polarized in the horizontal plane will interact with and propagate through hydrometeors differently than one polarized vertically. These differential observations are the basis of weather radar polarimetry. In addition to the dual-polarization of the transmitted signal (which can be alternate or simultaneous), polarimetric radar systems also have Doppler capability. A polarimetric radar is hence a dual-polarization Doppler radar system. Thus, while conventional radar measures only backscattered power, polarimetric radar measures additional quantities.

Status of Weather Radars in India

Radars are monitor by IMD for detection of various weather systems like, thunderstorms, hailstorm and tracking of cyclonic storms. They are also used in rainfall estimation and hail warning. Various meteorological, hydrological and aviation products generated from Doppler weather radar data using are extremely useful to the forecasters. Radars helps to estimate the storm's intensity, location and in forecasting its future path for safe navigation of aircrafts and ships. IMD has upgraded the conventional radars in the observational network with Doppler radars using digital technology. Indigenous efforts were also made to design and develop advanced Radar technology for cost effectiveness and self-reliance. Under an MOU with ISRO, DoS, one S-Band, Doppler weather radar was designed, developed and installed at Sriharikota in 2004. In 2007, under Modernization of IMDs observational network 13 old and obsolete analogue radars were replaced with Doppler Weather Radars (DWRs).

In second phase of Modernization plan of IMD, 30 DWRs are proposed to be procured and installed at various locations throughout India. There is a separate plan to install 9 DWRs at hilly region under the scheme "Integrated Himalayan Metrology Programme" for western and Central Himalayas. One X-band DWR is under installation at Srinagar to cater the needs of Amarnath Yatra. The existing Doppler weather radars have also been networked to provide data for numerical weather prediction models for now casting. Thus Radar network is playing significant role in the modernization of IMD's observational and forecasting systems.

Conclusion

Rainfall can be measured using rain gauges and radars. Raingauges provide accurate point measurement of rainfall, but, its spatial and temporal resolution are poor, as the density of raingauge stations is very less, especially in countries like India. Weather radars provide high spatial and temporal resolution of rainfall incidents. At the same time, the measurements are indirect and hence, caution is required to ward off inaccuracies. Accurate and reliable rainfall information is crucial for hydrological modeling and radar technology has the potential to provide such accurate information. Hence, radar rainfall data can be a valuable input for the estimation of hydrological responses such as runoff volume and peak flow.

REFERENCES

Babendreier J., Castleton K., 2005. Investigating uncertainty and sensitivity in integrated, multimedia environmental models: tools for FRAMES- 3MRA. Environmental, Modelling and Software 20: 1043–1055.

Brauer, C.C., Teuling, A.J., Overeem, A., van der Velde, Y., Hazenberg, P., Warmerdam,Dutch rainfall at hourly to daily scales. *Adv. Water Resour.* 45(1): 76–85

Guhathakurta P., and Rajeevan M. 2008. Trends in the rainfall pattern over India. *Int. J. Climatol.* 28: 1453–1469

Marani, M., 2005. Non-power-law-scale properties of rainfall in space and time. Water *Resour. Res.* 41 (8) http: //doi.wiley.com/10.1029/2004WR003822.

Moreau E, Testud J, Le Bouar E. 2009. Rainfall spatial variability observed by X-band weather radar and its implication for the accuracy of rainfall estimates *Advances in Water Resour.* 32: 1011–1019.

Moriasi D., Arnold J., Van Liew M., Bingner R., Harmel R., and Veith T. 2007. Model evaluation guidelines for systematic quantification of accuracy in watershed simulations. Transactions of the ASABE 50: 885–900

Nicótina, L., Alessi Celegon, E., Rinaldo, A., Marani, M., 2008. On the impact of rainfall

P.M.M., and Uijlenhoet, R., 2011. Anatomy of extraordinary rainfall and flash flood in a Dutch lowland catchment. *Hydrol. Earth Syst. Sci.* 15 (6): 568-582

Price, K., Purucker S., Kraemer S. and Babendreier J. 2012. Trade offs among calibration targets for watershed modeling. *Water Resour. Res.* 48: 10542,

Suresh R., 2010 Radar hydrology and rainfall mapping. Proceedings of the national seminar on Doppler Radar and Weather Surveillance (DRaWS-2010), Regional Meteorological Centre, Chennai, India

Reston, V.A., gongondo C. X., Gottschalk L. and Alemaw B. 2011. Evaluation of spatial and temporal characteristics of rainfall in Malawi: a case of data scarce region. Theoretical and Applied Climatology 106: 79–93.

Van de Beek, C.Z., Leijnse, H., Torfs, P., Uijlenhoet, R., 2012. Seasonal semi-variance of patterns on the hydrologic response. *Water Resour. Res. 256*(4): 214-224

Wilco, T.d., Leijnseb, H., Eertweghc, G., and Uijlenhoetd R., 2018 Spatial resolutions in areal rainfall estimation and their impact on hydrological simulations of a lowland catchment. J. Hydrol. 563(1): 319-335

Yang, D., Kanae S., Oki, T. and Musiake, K., 2001, Expanding the distributed hydrological modelling to continental scale. ln: Soil-Vegetation Atmosphere Transfer Schemes and Large-Scale Hydrological Models.270(1) : 125-134

Yang. D., Oki, T. Herath, S. and Mtisiake K., 2002. A geomorphology-based hydrological model and its applications. In: Mathematical Models of Small Watershed Hydrology and Applications. 9: 259-300.

Chapter 19

Effect of Pollution by Cement Industry in Trees and Crops

Sheelendra Kumar Upadhyay[1] *and Kuldeep Rajpoot*[2]

[1]*Assistant Professor, Department of Environmental Science, Manglayatan University, Jabalpur, Madhya Pradesh*

[2]*Research Scholar, Department of Agricultural Engineering, Institute of Agricultural Sciences, Banaras Hindu University, Varanasi, Uttar Pradesh*

Cement is the primary material for construction and civil engineering construction. Therefore, the cement manufacturing sector plays an important role in the economic development of the country. The purpose of this chapter is to show the environmental impacts of the cement industry, propose solutions, and improve the sustainability of the cement industry. The main environmental issues associated with cement production are the use of raw materials and energy as well as emissions into the air.

Introduction

Telecom, petroleum, coal, fertilizer, iron, steel, and cement are the major infrastructure sectors of the world. With the cement business additionally playing a major role, cement can result in a basic demand for all construction activities within a country's rapid ascent and growth. Cement is employed in housing, dams, bridges, industrial construction, roads *etc.* Therefore, cement is a basic material fully employed in the types of construction. Cement can be a finer material in which water forms a paste that gradually hardens. It is made by mixing mixed raw materials. Most of the material formed within the mixture silicates the metal element carbonates as sedimentary rock and various alumina, as clay or sedimentary rock. During the sintering process, a chemical process occurs, producing nodules, called clinkers that contain metal elements silicates and aluminates. Once the clinker is powdered with a small amount of mineral called Portland Cement. As cement manufacturers still search for various fuels, many of us wonder what type of fuel will be used in the combustion method. The typical energy input required to produce one ton of cement is 4 million BTUs - equivalent to about 389 pounds of coal.

Discovering the ways that each energy desires and reliance on fossil fuels can be a major priority for cement corporations. Although coal, crude oil coke, and various fossil fuels have historically been burned in cement kilns, many cement corporations have turned to various energy-rich fuels. Today, many plants use large forms of various fuels as a part of their overall energy plan; Their energy needs begin at ten to seventy per cent. Typically, shopkeepers from various industries are with efficiency burned as waste or by-product fuel. Occasional, cement can be a safe and well-tried exercise worth their energy. Coal and coke contain carbon and are burnt after heating in large quantities. However, coal and coke are not the only fuels that contain carbon. Tires are also an excellent supply of hydrocarbons (carbon and hydrogen). Victimization tires for combustion during cement ovens produce 25 per cent much more energy than coal, and can also cause fewer emissions. Any material with a high carbon content can be used as a fuel.

Paper, packaging, plastics, sawdust and solvents are all suitable to be used as various fuels. Due to extremely high temperatures (well over 3,000 degrees Fahrenheit), these materials burn quickly and very efficiently. Burning different types of fuel in cement kilns gives many environmental edges. Such energy recovery conserves valuable fossil fuels for future generations, while safely destroying waste that would accumulate in landslides well.

Cement Industry in the World

Since its introduction 2,000 years ago, there has been tremendous growth in cement production. While the use of cement in concrete has a long history (Malinowski, 1991), industrial production of cement began in the mid-19th century, first with shaft kilns, which were later used by rotary kilns as standard equipment worldwide. Was replaced. Today's annual global cement production reaches 2.8 billion tons and is expected to increase to some 4 billion tons per year in 2050 (Schneider *et al.*, 2011). Major development is expected in countries such as China and India, as well as regions such as the Middle East and North Africa. At the same time, the cement industry is facing challenges such as increasing costs in energy supplies (Lund, 2007), requirements to reduce CO_2 emissions, and the supply of raw materials in sufficient qualities and quantities (WBCSD, 2008).

Cement is used to make concrete for buildings and infrastructure, which are important for the quality of life and social and economic well-being. The cement sector is the third-largest industrial energy consumer, comprising 10.7 EJ, 7 per cent of global industrial energy use. Cement production involves the decomposition of limestone (calcium carbonate), which represents about two-thirds of the total CO_2 emissions generated in the process, with the remaining CO_2 emissions being due to fuel combustion. Thus despite substantial progress on energy efficiency, the use of alternative fuels and clinker replacements, the region has the second-largest share of 27 per cent (2.2 $GtCO_2$/yr) total direct 1 industrial CO_2 emissions in 2014. z Global Population and Urbanization. The pattern, coupled with the needs of infrastructure development, increases the demand for cement and concrete. Global cement production is set to increase by 12-23 per cent by 20-23 from the current level. Direct CO_2 emissions from the cement industry are expected to increase by 4 per cent globally by 2050 under the IEA Reference Technology Scenario (RTS), despite a 12 per cent increase in global cement production. Realizing the permanent change of the 2°C scenario (2DS) means a significant reduction of global direct CO_2 emissions by 24 per cent compared to current levels by 2050, with an expected increase in global cement production. This represents a cumulative emission reduction of 7.7 $GtCO_2$ compared to RTS by 2050, equivalent to approximately 90 per cent of current global industrial direct CO_2 emissions. z Adopting a whole life-cycle approach and working collaboratively along the entire construction value chain provides additional opportunities for carbon emission reduction beyond cement manufacturing. Optimizing the use of concrete in construction by minimizing waste, encouraging reuse and recycling, maximizing design life and utilizing the properties of concrete to minimize the operating energy of the built environment, leading to this sector There are strategies.

Cement Industry in India

India is the second-largest cement producer in the world and accounted for over 8 per cent of the global installed capacity as of 2019. India's overall cement production capacity was nearly 545 million tonnes (MT) in FY20. Of the total capacity, 98 per cent lies with the private sector and the rest with the public sector. The top 20 companies account for around 70 per cent of the total cement production in India. As India has a high quantity and quality of limestone deposits throughout the country, the cement industry promises huge growth potential. The demand of cement industry is expected to reach 550-600 MT per annum (MTPA) by 2025 because of the expanding demand of different sectors, *i.e.*, housing, commercial construction, and industrial construction.

According to CLSA (institutional brokerage and investment group), the Indian cement sector is witnessing improved demand. Key players reported by the company are ACC, Dalmia and Ultratech Cement. In the second quarter of FY21, Indian cement companies reported a sharp rebound in earnings and demand for the industry increased, driven by rural recovery. With the rural markets normalising, the demand outlook remained strong. For FY21, CLSA expects a 14 per cent YoY increase in EBITDA in the cement market for its coverage stocks. A total of 210 large cement plants account for a combined installed capacity of 410 MT in the country, whereas, 350 mini cement plants make up for the rest. Of the total 210 large cement plants in India, 77 are in the states of Andhra Pradesh, Rajasthan, and Tamil Nadu. Sale of cement in India stood at Rs 63,771 crore (US$ 9.05 billion) in FY20. India's export of cement, clinker and asbestos increased at a CAGR of 6.44 per cent between FY16-FY19. In FY20 (till January 2020), it reached US$ 1.66 billion. To enhance the source of capital for infrastructure financing, Credit Guarantee Enhancement Corporation, for which regulations have been notified by the RBI, will be set up in FY20.

According to the data released by Department for Promotion of Industry and Internal Trade (DPIIT), cement and gypsum products attracted Foreign Direct Investment (FDI) worth US$ 5.28 billion between April 2000 and March 2020. India's export of cement, clinker and asbestos increased at a CAGR of 1.68 per cent between FY16-FY20 and stood at US$ 1.98 billion in FY20. The Government of India is strongly focused on infrastructure development to boost economic growth and is aiming for 100 smart cities. The Government also intends to expand the capacity of railways and the facilities for handling and storage to ease the transportation of cement and reduce transportation cost. These measures would lead to increased construction activity, thereby boosting cement demand. As per Union Budget 2019-20, the Government expected to upgrade 1,25,000 km of road length over the next five years, which would boost the demand for cement. Also, the Government of India extended an additional outlay of Rs. 18,000 crore (US$ 2.43 billion) for the PM Awaas Yojana - Urban over the already allocated Rs. 8,000 crore (US$ 1.08 billion); this is expected to be used for the development of ~30 lakh houses (ground support for 12 lakh houses and completion of 18 lakh houses) and will likely create an additional 78 lakh jobs and boost production and sale in the steel and cement sectors.

Definition of Pollution

Pollution is the introduction of pollution into the natural environment that causes adverse changes. Pollution can take the form of chemical substances or energy, such as noise, heat or light. Pollutants, components of pollution, may be either foreign substances/energy or maybe naturally contaminated.

Types of Pollution by Cement Industry

Cement industries have been classified as highly polluting industries by the Central Pollution Control Board (CPCB). The main pollutants from the cement industries include particulate matter, sulfur dioxide (SO_2) and nitrogen dioxide (NO_2). Cement manufacturing contributes greenhouse gases directly through the production of carbon dioxide when calcium carbonate is heated (production of lime and carbon dioxide) and indirectly through the use of energy, especially if the energy is sourced from fossil fuels. it occurs. Effect of pollution of the cement industry on vegetable Stability of cement, concrete and cement replacement materials in construction M. Sombi,. p. Diederich, Sustainability of Construction Materials (2nd ed.), 2016.

Environmental Impact

Cement production is one of the largest industries in the world. Annual world production in 2013 was around 4 GT (about half of which was in China). It is produced in kilns of approximately 1400°C (2500°F), and about 750 kg (1650 lb) of CO_2 is released for each ton (2205 lb.). It is 400 m3 (524 yd3) gas, And CO_2 normally makes only 0.04 per cent of air. Cement production is responsible for about 7 per cent of greenhouse gas emissions worldwide. The only major contributors are power generation and transportation. Most of the CO_2 comes from chemical reactions when cement is made. The limestone from which it is made is calcium carbonate, and when it is heated it releases carbon dioxide and forms calcium oxide. Thus, while some benefit has been gained by making the process more energy-efficient, there is little scope for any further cuts.

Several different solutions to this problem have been proposed, including carbon capture and the invention of entirely new cement; But currently, the only effective way to achieve a significant reduction in greenhouse gas emissions is the use of cement replacement.

One process that has not been extensively investigated is carbon sequestration. In this process, hydrated cement reacts with CO_2 in the air, slowly reversing some of the processes in the kiln when the cement was made (this is a carbonation process; it also causes reinforcement corrosion, Section 25.3 See.2). It is estimated that this may reduce the carbon footprint of the cement industry by 3 to 5 per cent.

Reducing the Environmental Impact of Concrete

In Europe, cement production decreased by 26.9 per cent from 1990 to 2012, while CO_2 production decreased by 38.6 per cent, leading to an improvement in cement production (CEMBUREAU, 2014). However, to reach the objectives of various sustainability programs, further efforts should be made to improve every step in the concrete production line. For concrete, the main solutions to reduce the environmental impact of modern construction are (Flatt *et al.*, 2012):

- ✰ Partial Cement (Clinker) Replacement with SCMs
- ✰ Development of alternative binders
- ✰ Cement using concrete mix design limits the material
- ✰ Recycling of demolished concrete in new concave
- ✰ Sustainability (Designing new infrastructure for new service)
- ✰ Instead of demolishing and rebuilding the existing infrastructure (extending the service life of the existing infrastructure)

Some other technical solutions to improve cement production are also of interest, including CO_2 capture and storage. If this solution presents a major impact on the ultimate objective, it will require technological breakthroughs that are still in the research phase and are not planned to be ready for users before 2030, but they potentially account for 45 per cent of CO_2. Can be produced by cement (IEA and WBCSD, 2010). By 2050, base emissions will be 2.34 Gigatons (GT), but blue emissions of 1.55 Gt will be reduced by the contribution of energy efficiency (10 per cent), use of alternative fuels, and another fuel switching (24 per cent), a Kleiner replacement (10). per cent) And carbon capture and storage (56 per cent) (IEA and WBCSD, 2010).

Reducing Clinker Content

The worldwide use of mixed cement in the production of cement is already a general improvement, with significant investment for research conducted by cement producers. Many mixed cement types exist due to the great variability of usable SCM. These cement types are produced by replacing part of the clinker with SCM, which leads to cement with unchanged or improved properties for both general and special applications. The same qualities must be maintained; Otherwise, for a functional unit (concrete beam) and at the product scale (cement), more cement may be necessary to achieve the same durability. Therefore the beneficial effect of changing a part of the clinker may be useless (Li *et al.*, 2004). In the European context, binary cement comprises up to 35 per cent of each type of SCM, even in the case of inert mineral filler, and uses up to 80 per cent in large replacements of ternary cement clinker (EN 197-1, 2001). A final trend is the development of quaternary binder, but a lot of improvements are necessary to ensure a better understanding of the interaction between the clinker and the SCM.

The Government on Pollution by the Cement Industry

The Government is aware of the fact that cement industries are the major source of pollution in the country. The cement industries have been categorized as highly polluting industries by the Central Pollution Control Board (CPCB). The main pollutants emitted from cement industries include Particulate Matter, Sulphur Dioxide (SO2) and Nitrogen Dioxide (NO2). To prevent and control pollution from cement industries, CPCB conducts surprise inspection of cement industries under its Environmental Surveillance Squad programme to assess the compliance status concerning the prescribed emission/effluent norms. Directions under Section 5 of the Environment (Protection) Act (EPA), 1986 are issued to the non-compliant industries and under section 18(1)(b) of the Water (Prevention and Control of Pollution) Act, 1974 and the Air (Prevention and Control of Pollution) Act, 1981 to State Pollution Control Boards (SPCBs) to direct industries to comply with the stipulated norms. Further, the cement industries have also been directed to install on-line continuous (24x7) monitoring devices for better monitoring of compliance by CPCB and SPCBs. In August 2014 Ministry of Environment, Forest and Climate Change notified revised emission norms for Particulate Matter and also introduced emission norms for SO_2 and NO_2 for the cement industry. This information was given by Minister of State (Independent Charge) of Environment, Forest and Climate Change, Shri Prakash Javadekar, in Lok Sabha today.

REFERENCES

Acosta-Martínez V, Cruz L, Sotomayor-Ramírez D, Pérez-Alegría L (2007). Enzyme activities as affected by soil properties and land use in a tropical watershed. Appl. Soil Ecol., 35: 35-45.

https: //pubmed.ncbi.nlm.nih.gov/22367367/"bi.nlm.nih.gov/22367367/

India Brand Equity Foundation, Indian Cement Industry Analysis, Last updated on Jan, 5 202, https: //www.ibef.org/industry/cement-presentation#: ~: text=India per cent 20is per cent 20the per cent 20second per cent 20largest,the per cent 20rest per cent 20with per cent 20public per cent 20sector.

M. Sonebi, P. Diederich, Sustainability of cement, concrete and cement replacement materials in construction, Sustainability of Construction Materials (Second Edition), 2016

Mandre M (1995). Physiological and biochemical responses of coniferous trees to alkaline dust impact. In M. Mandre(Ed) Dust Pollution and Forest Ecosystem, A Study of Conifers in an Alkalized Environment. Inst. of Ecol. Pub. Tallinn, pp. 44-65

Peter A. Claisse, "Civil Engineering Materials, Chapter 17 Introduction to Cement and Material" https: //books.google.co.in/books?id=RtQsBgAAQBAJHYPERLINK "https: //books.google.co.in/books?id=RtQsBgAAQBAJ&pg" and HYPERLINK "https: //books.google.co.in/books?id=RtQsBgAAQBAJ&pg"pg

Peter A. Claisse, in Civil Engineering Materials, 2016

Press Information Bureau Government of India Ministry of Environment, Forest and Climate Change, "Pollution by Cement Industries", 11-August-2015 16: 33 IST https: //pib.gov.in/newsite/PrintRelease.aspx?relid=124810

Udara Arachchige, Janeesha Devasurendra, Limasha Chathumini, Kalani Alagiyawanna, "Environmental Pollution by Cement Industry", International Journal of Research 6(8): 631-635, July 2019G. Habert, Assessing the environmental impact of conventional and 'green' cement production, Eco-efficient Construction and Building Materials, 2014.

Chapter 20

Natural Regeneration of Tree Plantations: Impact on Soil and Microbial Characteristics, Insects and Spider Population

V. Davamani, E. Parameswari,*
R. Sangeetha Piriya and T. Ilakiya

Tamil Nadu Agricultural University, Coimbatore – 641 003, Tamil Nadu

The soil characteristics in the area of tree plantations plays a major role in the carbon sequestration potential of the trees and the living biomass, soil (roots and microbes), recalcitrant organic and inorganic carbon contributes. Carbon sequestration potential of tree plantations are affected by soil organic and inorganic carbon pools, carbon sequestration in tree biomass, rate of CO_2 evolution from the soil surface, carbon fluxes through net primary productivity. The acid concentration, degree of acid dissociation and the strength of acids present in the soil determines the soil pH. One of the most important ecological restoration techniques were the forest plantations using native- tree species. The forest succession might be catalyzed by plantations of native trees and also improves soil quality, facilities the recruitment of shrubs and trees and restores ecological interactions. In natural regeneration process, the species richness and density of natural regeneration were affected by the factors such as planted tree species, location of the plantation in the landscape and soil conditions. The impact of disturbance on insect diversity shows insect diversity would not much differ in undisturbed and disturbed area because of capability of insects by recolonizing in changed habitats. The role of microorganisms in maintaining the dynamic equilibrium and integrity of the biosphere is so critical that the continued existence of life is dependent on the sustained, microbial mediated transformation of matter in both terrestrial and aquatic environments. The basal respiration, dehydrogenase activity and metabolic coefficient are the indicators of microbial activity that influences the efficiency and oxidative capacity of the metabolic processes in microbes and also they were related to the quantity and quality of organic matter.

Impact of Tree Plantations on Soil Characteristics

The soil characteristics in the area of tree plantations plays a major role in the carbon sequestration potential of the trees and the living biomass, soil (roots and microbes), recalcitrant organic and inorganic carbon contributes for the carbon sequestration. From the research studies, it was found that carbon sequestration potential of tree plantations are affected by soil organic and inorganic carbon pools, carbon sequestration in tree biomass, rate of CO_2 evolution from the soil surface, carbon fluxes through net primary productivity. In the mixed plantation of *Acacia nilotica* and *Dalbergia sissoo*, the micro aggregates (250 micrometer to 53 micrometer) causes a formation of large fraction of soil aggregates that protected the soil organic carbon in the soil and were also related to soil stability. So, the soil organic carbon retention and protection against decomposition could be done by soil aggregates. But soil moisture, soil temperature and atmospheric temperature were significantly and positively correlated with the CO_2 rates that were evoluted from the soil surface (Arora and Chaudhry, 2017). Soil Inorganic Carbon at deeper layers and Soil organic carbon at surface layers has greater potential for carbon sequestration. Also, the studies suggested that the vegetation and soil that has large storage houses of carbon could play a significant role in decreasing or stabilizing the concentrations of atmospheric carbon dioxide which has been the mitigation of direct climate change consequences (Arora and Chaudhry, 2017). Carbon availability were used as biological indicators to estimate the soil quality (Knoepp *et al.*, 2000). The fine-root growth was significantly correlated with soil organic C, but not with their aboveground detrital inputs. And also the soil organic C had increased with the potential C mineralization which indicated that species influenced both the quality and quantity of SOC. In contrast to that, the increasing concentrations of fine-root lignin caused the decline in the SOC which revealed that the lignin derived C did not dominate SOC pools. The SOC causes effect on soil nutrient supply and soil physical and chemical properties so that the soil fertility was influenced by them. The relationships of soil fauna with SOC and bulk density were unknown (Russell *et al.*, 2007).

In *Tectona grandis* and *Albizia lebbeck* plantations, by the systematic soil sampling, it was found that the Organic carbon was significantly higher than the control plot. In addition to that, *Tectona grandis* plantation had lower organic levels than the *Albizia lebbeck* plantations as the lebbeck leaves contributed higher for the organic carbon and also, the pH and Bulk density of soil in both the plantations were found to be non-significantly lower than the control plot. And it was also revealed that the pH of the soil was almost neutral and slightly alkaline under *Tectona grandis* and *Albizia lebbeck* correspondingly (Imoro *et al.*, 2012). When comparing the plantations of *Q. suber* with the *E. globulus*, the bulk density of the soil in the upper horizon was found to be higher in the *E. globulus* plantations (Madeira, 1991).

The acid concentration, degree of acid dissociation and the strength of acids present in the soil determines the soil pH (Binkley *et al.*, 1989).With a high concentration of organic acids, low base saturation, the soil pH would be lower. Due to the aliphatic organic acids dissociation, the complex or chelate metals were formed that lowers the soil pH. Some of the strong acids such as oxalic, formic, citric and malic acids had a strong impact on soil pH (Zhang *et al.*, 2009). The pH of partially decomposed and humus layers of the pines were lower than the newly fallen litter (Rhoades, 1996). While maintaining the comparable plant diversity between the secondary forests and agroforestry, the agroforestry soil had possessed a higher pH (Sistla *et al.*, 2016).

Natural Regeneration Studies under Plantations

The plantation of native tree species plays a major role for ecological restoration, but the details of these processes that contribute for the ecological restoration was natural regeneration. Depending on initial species composition site conditions and plant density, the difference in the natural regeneration was observed in each of the native tree plantation that indicated the performance of plantation as a restoration energy (Sansevero *et al.*, 2011). And also, the forest regeneration in degraded lands were related to fruit dispersed limitation, seed bank losses, soil nutrient reduction, high seed predation and weed competition (Holl *et al.*, 2000; Zimmerman *et al.*, 2000). So, the ecological restoration techniques could be applied to restore community structure and ecological processes when the natural regeneration mechanisms were disturbed (Hobbs and Harris, 2001). One of the most important ecological restoration techniques were the forest plantations using native- tree species (Harrington and Edwards, 1999).

The forest succession might be catalyzed by plantations of native trees and also improves soil quality, facilities the recruitment of shrubs and trees and restores ecological interactions (Ruiz-Jaen and Aide, 2005). In the process of natural regeneration, the species richness and density of natural regeneration were affected by the factors such as planted tree species, location of the plantation in the landscape and soil conditions (White *et al.*, 2004). In the five mixed plantations of native tree established in abandoned pastured in the Brazilian Atlantic forest, it was found that there was an intense vegetation (naturally regenerated) over a period of 10 years (Sansevero *et al.*, 2011). In tropical forest, regular flowering and fruiting occurs which could be annually predicted but declined due resource limitations. In tropical forest, most of the early successional plants and other species reproduce by flowering earlier in their lives and annually they produce fruits, by this way regenerate.

For Seedling germination, rainfall and temperature are most crucial (Yadav and Sharma, 1968). Apart from rainfall and temperature, soil nutrient plays a vital role and influence regeneration. (Yadav, 1963) examined that sal tree exhibited lower natural regeneration when there was high nutrient status of Mg, P, K and exchangeable Ca in that soil. In addition to that, there was a low plant mortality rates up to the third year of planting, in three of the native tree plantations. Then, the high mortality rates were observed due to the natural consequence of the shorter lifetimes of pioneer species (Budowski, 1965). Though the plant recruitment was facilitated by tree mortality (Canopy openness), the colonization by weeds had occurred in the areas that were mostly surrounded by pastures and that limited tree and shrub recruitment (De Souza and Batista, 2004). Because of the high abundance of *C. pachystachya* (Vrtiaceae), it might be suggested that it could be established after one year of planting. Further, the recruitment of woody plants might be enhanced by eliminating the weeds (Holl *et al.*, 2000).

In the 28 years old forest plantations, 69 per cent of the individuals were derived from the natural regeneration (Pulitano *et al.*, 2004). And the plantations might affect the understory recruitment by changing environmental conditions and by the attraction of dispersal agents that caused low floristic similarity between the layers. In addition to that, a closed canopy was formed due to the native- tree plantations could alter the environmental conditions by the inhibition of settlement of plant pioneer species (Menalled *et al.*, 1998). Moreover, the species composition and the over story structure could improve seed rain and subsequently affect the species composition of naturally regeneration vegetation (Mesquita *et al.*, 2001).

It was also found that some species were able to develop under both sun and shade conditions (Eg: *Guarea guidonia*) (Parrotta, 1995). And that was due to the physiological adaptability to certain environmental changes, mainly light and also the temporary changes in the light entry caused by the canopy of deciduous plants (Gandolfi *et al.*, 2007). As there was a recruitment and persistence of *Guarea guidonia* in the understory, that was contributed due to the deciduous species such as *Cithasexylum myrianthum* (Verbenaceae), *Pseudobombax grandiflorum* (Malvaeceae), and *Sparattosperma leucanthum* (Bigniniaceae) (Russell *et al.*, 2007). Some of the other factors that have contributed to the high recruitment of Guarea guidonia were its fruit dispersal by frugivores birds and high fruit production (Carlo *et al.*, 2003).

There was a predominance of zoochorous species in the natural vegetation that followed the well-known tropical forest pattern (Jordano, 2009). In addition to that, the increased attractions of forest plantations of frugivores might increase the dispersal of propagules into site (Wunderle, 1997). The isolated zoochorous trees in pastures were able to receive a high seed rain beneath their canopies (Vieisa *et al.*, 1994). For the acceleration of ecological succession, the inclusions of zoochorous trees in restoration projects were considered as an important strategy (Reay and Norton, 1999). In another scenario, the small numbers of species with larger-sized propagules in the natural regeneration might be contributed by the low abundance of large mammals in an area (Galetti *et al.*, 2006).

Insect and Spider Diversity in Different Plantations

Mathew *et al.* (1998) studied diversity and abundance of insects four parts of western ghats, Kerala *viz.*, Silent valley, Nelliyampathy, Sholayar and Parambikulam. Altogether 1250 species of insets belonging to 15 insect orders were recorded and maximum numer of species belonged to Lepidoptera and Coleoptera. Their study shows Silent Valley contained a higher number of individuals and species (10451/578). The impact of disturbance on insect diversity shows insect diversity would not much differ in undisturbed and disturbed area because of capability of insects by recolonizing in changed habitats.

U Usha A (2015) documented insect diversity in Wadakkanchery thick vegetation hilly area and in some plain land. The insect orders prevalent in the study area are Coleoptera, Orthoptera, Diptera, Hymenoptera, Lepidoptera, Hemiptera,

Dictyoptera, Phasmida and Odonata. Relative abundance of species were high in order Coleoptera (30.82) and least in the order Phasmida (0.56).

Ngongolo *et al.* (2015) studied diversity and abundance of insects in Kichi forest, Tanzania and in adjacent local communities farms. Kichi forest showed higher abundance and farm land showed least abundance. Species like Aspidimorpha sp, Cophogryllus sp and Streblognatus sp was observed in both sites.

Kyerematen *et al.* (2014) at Muni – Pomadze Ramsar Site, Ghana reported a total of 134 families from 19 insect orders. Yenku Block A recorded the highest species richness (98) and the highest diversity index (14.97) with a relatively even distribution of species (0.834).

R *et al.* (2013) at Kota Damansara Community Forest Reserve, Malaysia reported 13 insect orders and 79 families. Among them Coleoptera (42.63 per cent) shows high richness followed by Hymenoptera (17.6 per cent), Diptera (10.8 per cent) and Orthoptera (10.85 per cent). They also showed more number of insects were collected by light trap (21 per cent) comparing malaise trap (12 per cent) and pitfall trap (4 per cent). They recorded more adundance of insects in the Kembit zone than other zones in the forest.

Jose *et al.* (2018) conducted a study to document the diversity of spider fauna inhabiting in the different ecosystems of Kavvayi river basins. The present study resulted in the documentation of 112 species of spiders belonging to 81 genera and 21 families. Araneidae was the most dominant family which constitutes 21.5 per cent of the total spider species collected. The second dominant family was Salticidae which constitutes 19.5 per cent of total spider population. Guild structure analysis of the collected spiders revealed seven feeding guilds, namely stalkers, orb web builders, ambushers, foliage runners, space web builders, ground runners and wandering sheet weavers.

Palem *et al.* (2016) studied diversity of spider fauna (Arachnida) in different ecosystems, Eastern ghats, Southern Andhra Pradesh of four different areas *viz.*, Sri Lankamala wildlife Sanctuary, (Kadapa district), Rapur Ghat forest, (Nellore district), Pallakondalu, (Kadapa and Chittoor district), Seshachalam Biosphere Reserve forest (Kadapa, Chittoor and Nellore district). During their study a total of 19, 25, 31 and 41 spider taxa belonging to 16 familes were recorded. In this Seshachalam Biosphere Reserve forest shows higher dominance index of 0.04213, Shanon (3.921) and Simpson diversity index (1.231).

Shirbhate and Shirbhate (2017) studied diversity and distribution of spider fauna (Family- Araneidae) in and around the Katepurna Sanctuary, Akola India. A total 11 genus and 26 species of family Araenidae (Arachnida: Araneae) were recorded. The generic diversity is in the order of Neoscona (9), Cyclosa (3), Larinia (3), Cyrtophora (2), Eriovixia (2), Poltys (2), Araneus (1), Argiope (1), Chorizopes (1), Gasteracantha (1), Zygiella (1).

Soil Microorganisms

The biological properties of the soil are dictated by the macro and microorganisms. Good physical and chemical properties supply the right environment and sufficient nutrients to the organisms for optimal biological activity. Soil microbial population is significant because it is often recognized as an important index of the health of soil ecosystem (Blaodatsky and Richter, 1998). The role of microorganisms in maintaining the dynamic equilibrium and integrity of the biosphere is so critical that the continued existence of life is dependent on the sustained, microbial mediated transformation of matter in both terrestrial and aquatic environments (Harris, 2003) and Hussain *et al.*, 2015).

The basal respiration, dehydrogenase activity and metabolic coefficient are the indicators of microbial activity that influences the efficiency and oxidative capacity of the metabolic processes in microbes and also they were related to the quantity and quality of organic matter (Bini *et al.*, 2013). Generally, the pH in pure plantations were significantly lower than the mixed plantations. In Chinese plantations, there was a significantly negative correlation between pH and C:N ratio which concluded that the bacteria might occur in soils with lower acidity than fungi. The ratio of respired Carbon to biomass Carbon was termed as Microbial metabolic quotient and that was considered as an index for evaluating substrate utilization efficiency of soil microbial community and also the soil pH and microbial metabolic quotient were negatively related in the Chinese fir plantations. Also, the more CO_2-C per unit biomass and per unit time were produced by microbial communities under pine forest soil conditions than communities at a more neutral pH range (Qing-Kui *et al.*, 2008).

REFERENCES

Arora, P., Chaudhry, S., 2017. Vegetation and soil carbon pools of mixed plantation of Acacia nilotica and dalbergia sissoo under social forestry scheme in Kurukshetra, India. J. Mater. Environ. Sci. 8. https: //doi.org/10.26872/jmes.2017.8.12.482

Bini, D., Santos, C.A. dos, Bouillet, J.P., Gonçalves, J.L. de M., Cardoso, E.J.B.N., 2013. Eucalyptus grandis and Acacia mangium in monoculture and intercropped plantations: Evolution of soil and litter microbial and chemical attributes during early stages of plant development. Appl. Soil Ecol. 63, 57–66. https: //doi.org/10.1016/j.apsoil.2012.09.012

Binkley, D., Valentine, D., Wells, C., Valentine, U., 1989. An empirical analysis of the factors contributing to 20-year decrease in soil pH in an old-field plantation of loblolly pine. Biogeochemistry 8, 39–54. https: //doi.org/10.1007/BF02180166

Budowski, G., 1965. Distribution of tropical American rain-forest species in the light of successional processes.

Carlo, T.A., Collazo, J.A., Groom, M.J., 2003. Avian fruit preferences across a Puerto Rican forested landscape: Pattern consistency and implications for seed removal. Oecologia 134, 119–131. https: //doi.org/10.1007/s00442-002-1087-1

De Souza, F.M., Batista, J.L.F., 2004. Restoration of seasonal semideciduous forests in Brazil: Influence of age and restoration design on forest structure. For. Ecol. Manage. 191, 185–200. https: //doi.org/10.1016/j.foreco.2003.12.006

Gandolfi, S., Joly, C.A., Rodrigues, R.R., 2007. Permeability - Impermeability: Canopy trees as biodiversity filters. Sci. Agric. https: //doi.org/10.1590/ S0103-90162007000400015

Harrington, T.B., Edwards, M.B., 1999. Understory vegetation, resource availability, and litterfall responses to pine thinning and woody vegetation control in longleaf pine plantations. Can. J. For. Res. 29 1055-1064.

Harris, J.A., 2003. Measurements of the soil microbial community for estimating the success of restoration. Eur. J. Soil Sci. 54, 801–808. https: //doi.org/10.1046/ j.1351-0754.2003.0559.x

Hobbs, R.J., Harris, J.A., 2001. Restoration ecology: Repairing the earth's ecosystems in the new millennium. Restor. Ecol. 9, 239–246. https: //doi.org/10.1046/ j.1526-100X.2001.009002239.x

Holl, K.D., Loik, M.E., Lin, E.H.V., Samuels, I.A., 2000. Tropical montane forest restoration in Costa Rica: Overcoming barriers to dispersal and establishment. Restor. Ecol. 8, 339–349. https: //doi.org/10.1046/j.1526-100X.2000.80049.x

Imoro, A.M., Sackey, I., Abubakari, A., 2012. Preliminary Study on the Effects of Two Different Sources of Organic Manure on the Growth Performance of Moringa oleifera Seedlings. academia.edu 2, 147–158.

Jordano, P., 2009. Fruits and frugivory., in: Seeds: The Ecology of Regeneration in Plant Communities. CABI, pp. 125–165. https: //doi.org/10.1079/9780851994321.0125

Jose, A.C., Sudhin, P.P., Prasad, P.M., Sreejith, K.A., 2018. Spider Diversity in Kavvayi River Basin, Keralà, Southern India. Curr. World Environ. 13, 100–112. https: //doi.org/10.12944/cwe.13.1.10

Knoepp, J.D., Coleman, D.C., Crossley, D.A., Clark, J.S., 2000. Biological indices of soil quality: An ecosystem case study of their use. For. Ecol. Manage. 138, 357–368. https: //doi.org/10.1016/S0378-1127(00)00424-2

Kyerematen, R., Acquah-Lamptey, D., Owusu, E.H., Anderson, R.S., Ntiamoa-Baidu, Y., 2014. Insect Diversity of the Muni-Pomadze Ramsar Site: An Important Site for Biodiversity Conservation in Ghana. J. Insects 2014, 1–11. https: //doi. org/10.1155/2014/985684

Madeira, M.A.V., 1991. Influence of mineral nutrients redistribution in eucalyptus plantations on soil properties, in: Developments in Geochemistry. Elsevier, pp. 485–494. https: //doi.org/10.1016/B978-0-444-88900-3.50052-7

Menalled, F.D., Kelty, M.J., Ewel, J.J., 1998. Canopy development in tropical tree plantations: A comparison of species mixtures and monocultures. For. Ecol. Manage. 104, 249–263. https: //doi.org/10.1016/S0378-1127(97)00255-7

Ngongolo, K., Mtoka, S., Rubanza, C.D., 2015. Floral Visitors and Pollinators of Sesame (Sesamum indicum L) from Kichi Forest to the Adjacent Local Communities' Farms. Entomol. Appl. Sci. Lett. 2, 32–39.

Palem, H., Kanike, S., Purushottam, V.R.S., 2016. image: https: //crossmark.crossref. org/images/crossmark_button.png Diversity of Spider Fauna (Arachnida:

Araneae) in Different Ecosystems, Eastern Ghats, Southern Andhra Pradesh, India. South Asian J. Life Sci. 4, 51–60. https: //doi.org/10.14737/journal.sajls/2016/4.2.51.60

Parrotta, J.A., 1995. Influence of overstory composition on understory colonization by native species in plantations on a degraded tropical site. J. Veg. Sci. 6, 627–636. https: //doi.org/10.2307/3236433

Pulitano, F.M., Durigan, G., Dias, L.E., 2004. A Mata Ciliar da Fazenda Cananéia: Estrutura e Composição Florística em Dois Setores com Idades Diferentes. Pesqui. em Conserv. e Recuper. Ambient. no Oeste Paul. (resultado da Coop. Bras. 419–446.

R, K.A., A, A.A., R, M.S., 2013. Diversity and abundance of insect species at Kota Damansara Community Forest Reserve, Selangor. Sci. Res. Essays 8, 359–374. https: //doi.org/10.5897/SRE12.481

Reay, S.D., Norton, D.A., 1999. Assessing the success of restoration plantings in a temperate New Zealand forest. Restor. Ecol. 7, 298–308. https: //doi.org/10.1046/j.1526-100X.1999.72023.x

Rhoades, C.C., 1996. Single-tree influences on soil properties in agroforestry: Lessons from natural forest and savanna ecosystems. Agrofor. Syst. 35, 71–94. https: //doi.org/10.1007/BF02345330

Ruiz-Jaen, M.C., Aide, T.M., 2005. Restoration success: How is it being measured? Restor. Ecol. https: //doi.org/10.1111/j.1526-100X.2005.00072.x

Russell, A.E., Raich, J.W., Valverde-Barrantes, O.J., Fisher, R.F., 2007. Tree Species Effects on Soil Properties in Experimental Plantations in Tropical Moist Forest. Soil Sci. Soc. Am. J. 71, 1389–1397. https: //doi.org/10.2136/sssaj2006.0069

Sansevero, J.B.B., Prieto, P.V., de Moraes, L.F.D., Rodrigues, P.J.F.P., 2011. Natural regeneration in plantations of native trees in lowland Brazilian Atlantic forest: Community structure, diversity, and dispersal syndromes. Restor. Ecol. 19, 379–389. https: //doi.org/10.1111/j.1526-100X.2009.00556.x

Shirbhate, M. V., Shirbhate, A.M., 2017. Diversity and distribution of Spider fauna (family- Araneidae) in and around Katepurna Sanctuary, Akola, India. Environ. Conserv. J. 18, 45–52. https: //doi.org/10.36953/ecj.2017.18306

Sistla, S.A., Roddy, A.B., Williams, N.E., Kramer, D.B., Stevens, K., Allison, S.D., 2016. Agroforestry practices promote biodiversity and natural resource diversity in atlantic Nicaragua. PLoS One 11. https: //doi.org/10.1371/journal.pone.0162529

U Usha A, J.V.K., 2015. A study on insect diversity of a selected area in Wadakkanchery (Thrissur, Kerala). J. Zool. Stud. Journalofzoology.com J. Zool. Stud. JOZS 2, 38–50.

Wunderle, J.M., 1997. The role of animal seed dispersal in accelerating native forest regeneration on degraded tropical lands, in: Forest Ecology and Management. pp. 223–235. https: //doi.org/10.1016/S0378-1127(97)00208-9

Yadav, J., 1963. Studies on soil profiles in Chakrata division of Uttar Pradesh. Indian For. 89, 18–38.

Yadav, J., Sharma, D., 1968. A Soil Investigation with Reference to Distriubution of Sal and Teak in Madhya Pradesh. undefined.

Zimmerman, J.K., Pascare, J.B., Mitchell Aide, T., 2000. Barriers to forest regeneration in an abandoned pasture in Puerto Rico. Restor. Ecol. 8, 350–360. https: //doi. org/10.1046/j.1526-100X.2000.80050.x

Index

D

E

F

G

Q

R

S

T